EVERYTHING IS
MATTER
MOVING
THROUGH SPACE

JOSEPH PALAZZO

authorHOUSE®

AuthorHouse™
1663 Liberty Drive
Bloomington, IN 47403
www.authorhouse.com
Phone: 1 (800) 839-8640

Published by AuthorHouse   10/10/2018

ISBN: 978-1-5462-5619-9 (sc)
ISBN: 978-1-5462-5618-2 (e)

Library of Congress Control Number: 2018909800

Print information available on the last page.

Contents

Preface

One of my wishes is that one day a reader will seize upon the ideas presented in this book and launch the next revolution so desperately needed in physics. My voice is a lonely one. Once there were hundreds of voices competing for your attention. Now in the age of the internet, there are billions. And so a message can easily be drowned out by a tsunami of voices. But if the message contains a kernel of undeniable truth then it is worth the time and energy in disseminating it to the greater public.

There is another reality we must all confront: the real language of physics is mathematics. That is incontrovertible. And in the past, the verbal interpretation of these equations in physics were too often exaggerations, the kind which bordered on the fantastic. It might have been the thing to do in order to capture the public imagination, and the needed funding. However there is a danger in that sort of enterprise, which makes it difficult to walk away. My hope is for this book to be a cautionary warning to the next generation: learn the math, filter out the verbal exaggerations, and aim to restore some sense of logical order in this great endeavor. The rewards will be immense.

But let's move on. Chapter 1 on the notion of time reveals to what point we have misused that concept. The time taken to mull over this notion was well spent as a breakthrough occurred when I realized what should be the answer to the question - how do we measure motion? Surprisingly, the answer is, **with motion**. The trick was to reverse conventional thinking, which has always been that motion is measured in terms of the ratio distance over time, and therefore time had to be considered as a fundamental concept, and motion, as a derived concept. Why is a reversal of the conventional thinking necessary? Because motion is that which is observed, while time is a mental construct. Because in a universe with no motion, time is a useless concept. A clock is a simple device with **internal moving parts** that conveniently facilitates the measuring of motion. You need motion to measure motion – this was the major breakthrough. Unquestionably then time is one of the greatest inventions the human mind has ever produced.

Even though a new thinking about time won't necessarily change any of the fundamental equations of physics already established, it brings a new perspective and is an invitation to revise our old notion of time, particularly in regard to the question: is time real or is it an illusion? As Einstein once put it: "The separation between past, present and future is only an illusion, although a convincing one."

Chapter 2 is about the real nature of a Minkowski diagram. With this new concept of time, a Minkowski diagram, which has been taken so far as a coordinate system, is reduced to the status of a graph. The consequence is that we do not live in a 4-D world, but simply in a 3-D with time acting as a parameter. The 4-vector formalism is a convenient way to deal with Lorentz invariance.

Chapter 3 outlines three laws of kinematic, emphasizing the 3rd law as the underpinning of the 2nd law of thermodynamics, namely entropy. It challenges our notion of the vacuum energy, and it puts into question the reality of Hawking radiation. Its impact on thermodynamics gives rise to the question: why is energy quantized?

Chapter 4 revisits the old debate between Einstein and Bohr on the meaning of Quantum Mechanics, giving new clarity into the postulates of Quantum Mechanics.

The primary focus of Chapter 5 is on the claim that General Relativity is not a theory of gravity, even though gravity plays in it a pivotal role. Gravity is then seen as a fictitious force such that in a strong gravitational field its effects are seen in the bending of light and the anomaly of objects in a strong gravitational field such as the orbit of Mercury.

Chapter 6 is an examination into the fundamental reason why gauge theory in GR Is different than in QFT, and why that matters in the nearly impossible task of quantizing GR.

Chapter 7 offers a simple explanation of the Cosmic Microwave Background, regardless of any cosmological model, and re-examines the questionable assumptions of the Big Bang Theory.

Last, but not the least, chapter 8 is a revision on why String Theory fails to be what everyone anticipated to be, the Theory of Everything.

The title is the unifying theme for all of these disparate subjects.

Joseph Palazzo

"Mathematics is the art of giving the same name to different things"

- Henri Poincaré (1854-1912)

Chapter 1

New Insights into the Concept of Time

(Thinking is mapping)

It's been the conventional thinking in physics that time is a fundamental concept, and motion, a derived one. The proposal in this chapter is that motion is the fundamental concept. We do not move in time, except perhaps figuratively. In the real world, matter is simply in motion. Time is the concept needed to measure that motion.

Does time exist? It exists as an invention of the human mind, just like the number system and the alphabet – just to mention two extraordinary inventions of the human mind.

What we experience through the passing of the seasons is the earth's trek around the sun. The cycle of daytime yielding to nighttime to go back again to daytime is the earth rotating on its own axis. When you look at a clock, you see the needle moving from tick to tick. Or the swinging of the pendulum in a grandfather clock. A digital clock uses the oscillations of a crystal to keep time. So what is involved directly or indirectly in all those instances is the motion of a body. What is real is "matter moving through space". And that should be the first hint: motion, and not time, is fundamental.

You might argue: "But there are changes occurring in nature that doesn't involve motion – take the leaves changing colors in the fall." Those changes occur because of changes in the energy coming from our sun. But what is energy? It's the ability to do work, and what is work? It's the ability to apply a force through a distance. And what is that force doing? It's "moving matter through space". In Quantum Field Theory, the notion of "force" is replaced by "an exchange of a force-mediating particle", which is still "matter moving through space".

In the zillions of concepts we have conjured in our minds, there are those that have been quite reliable – besides energy, momentum and entropy comes to mind. And time is such a concept that falls into this category. Such concepts are so crucial in the toolkit of our knowledge that without them it would be nearly impossible to understand the workings of our universe that we often think of them as fundamental. But everything that we see in the real world is "matter moving through space", and that includes all of your thought processes.

To explore the nature of time furthermore we must delve into the nature of mappings as it is not only fundamental to the structure of mathematics but more importantly it is also fundamental in our ways of thinking. But first we start with how we are going to use logic in our deliberations.

1.1 Assumptions, Logic and Worldview

The logic we have in mind consists of three major steps, denoted as: (1) unprovable truths (2) logic concerning truths derived from unprovable truths, and (3) stretching the logic to its extreme limits.

This logic is based on Gödel's incomplete theorem: "Any consistent formal system F within which a certain amount of elementary arithmetic can be carried out is incomplete; i.e., there are statements of the language of F which can neither be proved nor disproved in F."

The incomplete theorem contains two of the three steps cited above: (1) "there are statements of the language of F which can neither be proved nor disproved in F", (unprovable truth); (2) "formal system F within which a certain amount of elementary arithmetic can be carried out", (logic concerning provable truths). We now stretch this to its extreme limits (3) by considering F to be any system that a human mind can think of. This includes all spheres of human endeavor: not only math but language, science, philosophy, art, politics, the social sciences, the humanities, culture, etc. There are no exceptions. The irrational can be construed as logical inconsistencies.

By unprovable truths we mean axioms in math or hypotheses in science, to name two instances. These often go in the language of everyday by the name of assumptions. They are the starting points from which other concepts are derived using some form of logic. It is imperative to know these assumptions, that is, getting familiar with their meaning and applicability. Many assumptions often carry their own hidden assumptions, sometimes in the form of special conditions or limited circumstances. It's also important to realize that unknown assumptions might very well underlie the very assumptions we hold onto, and so one must be open and vigilant that any of these can arise at any time.

By logic we mean all informal, formal, symbolic and mathematical logic. Since many books have been written on this, we will not delve too deeply into this subject, but only use what we need. The basic importance of logic is that it yields consistency, that is, it has the ability of filtering out inconsistencies and contradictions.

By stretching the logic to its extreme, we've have already done that by stretching Gödel's incomplete theorem from a mathematical system to all system of human thoughts.

In the well-known Epicurean paradox, one version is: if God knows everything, including the future, then everything is predetermined and there is no free will (stretching the logic). This brings out the following question: What do you make of all the evil in the world? Such paradox forces one to re-examine one's own assumptions. Regardless of the solutions that were proposed to this riddle in 2300 years since it was articulated by Epicurus, no one has found a satisfactory resolution because this leads to a deity responsible for all the evil and suffering in this world, contradicting the notion that God is all-good (another assumption).

A worldview not only consists of assumptions and a logic based on those assumptions but also determines one's moral compass, one's priorities in life, one's core values, one's attitude, in particular towards empirical evidence, in other words, one's general philosophical position.

We will use the symbol "→" to mean a "mapping" in the most general sense. "Mapping" in the narrow mathematical sense, which is slightly more precise, will be in use later on.

The above statements can be stated as follow:

Assumptions → core beliefs → worldview

Note: Logic is bookkeeping for consistency. But it is pre-conditioned on one's assumptions. The whole package is part of a worldview.

In establishing concepts, and by "a concept" we mean "a mental construct", we have,

Assumptions → theories (concepts derived with logic)

At one point, we must return to the real world, that is,

Theory → predictions

Where predictions are observables in the real world. Without this last step, we have a theory but not a scientific one.

In the world of science, we then have,

Scientific theory → predictions → real world

The last stretch can only be accomplished by empirical investigation, that is, a theory that makes prediction of unknown facts, which are subsequently verified in the real world.

Science has the responsibility to close the loop.

Real world → concepts → assumptions → scientific theory → predictions → real world

For the rest of this book, a "theory" will stand in for a "scientific theory."

1.2 Mapping

A unique ability humans have is that every thought the human mind can hold in his or her brain can be written as a mapping.

Another fundamental ability of the human brain is to form sets along with mappings between various sets. Humans have been doing this since the beginning of time but only became aware of this process not that long ago, relatively speaking.

Math and language are the results of this unique ability.

At the base of this structure, our mind does the following function:

Real world → a mental construct

For instance,

Bird (observation in the real world) → the word "bird" (in the English language)

This mapping is more sophisticated than what it seems at first look. At a more primitive level, we have:

The sound "bird" → the word "bird"

This can be symbolized as:

Observations in the real world → sounds → symbols
(words/sign language for those unable to hear)

Not to confuse the above mapping with its reverse:

A mental construct → real world

In math, the reversal of an arrow is the inverse mapping. But we're not doing a mapping between two sets in the mathematical sense. This will come later. In the above case, one would read the statement to mean that as one invents an idea then one looks for something in the real world that would correspond to that idea. Think of creatures such as leprechauns, fire-breathing dragons, and werewolves.

Imaginary creatures → real world

We have to insist on the "imaginary" part of this mapping until evidence shows to the contrary.

For now, it's just sufficient to say that we will examine mappings going in this direction,

Real world → a mental construct

That is, the process we are alluding to initiates from the real world and it is then mapped out in our brain as a mental construct. This is how we teach language to our young ones: by pointing to an object "bird" and repeating the word/sound "bird". It's the most basic thing that human brains can do – a mapping between the real world and a symbol that our brain can grapple with, remember, assimilate and reproduce at some later times - a joy expressed by parents eager to recount their child's first spoken word.

Once that process has taken place, nothing forbids the child's versatile brain to quickly jumpstart to a realm of the imagination, which is fine as we often need creativity more often than reason in order to invent new mappings, new mental constructs, and so on.

When one of the side of a mapping is made up of elements of the real world, such as a bird, which I can point to with my fingers, this is the closest we can get to a true statement. Should I be pointing to something other than a bird, you would know that my statement, "This is a bird", would be false. Such mappings with one side of the mapping pertaining to the real world are of the 1st degree of abstraction.

Another example of a mapping of the 1St degree of abstraction, taking cues from physics, would be:

Motion (observed in the real world) →

Vector representation of a velocity

When the two sides of the mapping are themselves abstractions, we are at a higher level of abstraction (2nd degree). For instance,

A quantum state → a vector in a Hilbert space

Here vectors in a Hilbert space can take any dimension from one to infinity (as opposed to 3 for motion), and do not necessarily represent an object in the real world with a sense of direction like motion (a quantum state is such an object). These vectors in a Hilbert space obey certain mathematical rules, which are the same mathematical operations of ordinary vectors for motion. Such mappings of 2nd degree are more abstract, harder to establish and can be difficult to deal with. But in our quest to understand the universe, they are inescapable.

1.3 Math

Math is a powerful tool mainly because it is self-referential: it is based on the law of identity, of which mapping is the most powerful. On the other hand, physics has an additional component, which math is free from: physics must describe the real world, and hence it must have the ability to make predictions. There is more hand-waving than one would wish for, but this is inevitable as math, being a mental construct, within its creative process, is boundless. But It does have its own limitations in describing the real world. The real task is for science to determine which ideas we are so fond of creating really describe the real world. The mathematicians often straddle the world of science and the world of art as their work is descriptive in nature. Physicists, on the other hand, are more constrained in their creativity as their theory must explain not just any facts but how they pertain to the explanation of the mysteries of the universe.

Math is as stated before a mental construct. Consider these two descriptions: (a) I see a sheep in the field in front of me, I see another sheep, and then another one; or (b) I see three sheep in the field in front of me.

Notice that description (b) is shorter and easier to handle. Were I to count an additional sheep, description (b) is easily updated by saying: now I see four sheep. This particular description lends itself to mathematical manipulations and the richness it entails.

Nevertheless numbers just like words are descriptions. Now a number system is only possible if I have already drawn a set in my mind. Before I count the windows in this room, I need to form the set of "all the windows in this room". Otherwise I might be counting the bricks in the wall, the doors, the frames on the wall, etc. In the above example, the set was "the sheep in the field in front of me". I'm not counting the sheep in someone else's field. If I did, I would have to enlarge my set: "all the fields surrounding me" for instance would be another example.

Necessarily the number system, just like the alphabet system, is a mental construct. A very important development was when we put numbers to a stick, and call that "measuring the length of the stick "– we had just invented, mentally speaking, a space connected with geometry (the stick is a line). When we extended that concept to areas – one stick on the horizontal and another stick on the vertical - this completed the "square" and in so doing we also discover the irrational number $\sqrt{2}$. The imaginary number i $= \sqrt{-1}$ was initially invented to provide a solution to the equation $x^2 + 1 = 0$. We then discover that rotations are perfectly fitted to be represented by the invention of these imaginary numbers.

Note: the words "discover" and "invent" are intimately connected: we "invent" a concept, then go on an exploration of that idea, and then "discover" some qualities initially unsuspected in our invention. For instance after we <u>invent</u> a number system, we <u>discover</u> that some numbers are divisible by the number two, and there are those numbers that aren't. We then <u>invent</u> the concept of "even numbers" for the first case, and the "odd numbers" for the second case, which can lead to further discoveries and inventions.

So now we will define the word "mapping" in its narrow mathematical sense (an invention). Math is then the manipulations of mappings (a discovery).

Consider any two sets M and N, and a mapping f:

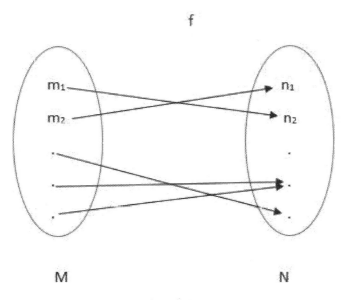

M   N

Fig. 1.1

Symbolically, this is written as:

$$f: M \rightarrow N$$

This means that each element of the set M is mapped onto an element of the set N. Note that not all the elements of N have a corresponding element in M.

One can choose an infinite number of sets like M, an infinite number of sets like N, and an infinite number of mappings like f between them. The above illustration is one example of that infinite soup. Now, out of this infinite soup, there is a lot of junk. Mathematicians are in the business of filtering out the good stuff from that junk. They identify different types of mappings (one-to-one, many-to-one, "onto", "into", injective, surjective, and so on). They also identify different sets (natural numbers, integers, rational numbers, etc.). They also narrow down certain sets to yield topologies, manifolds, bundle fibers, etc. They further narrow down certain mappings to give smooth manifolds, linearity, differentiability, etc. In this infinite soup that we've created, we can discover a number system that deals with the square root of minus one, and discover its properties. Math is no doubt a very creative process: within its logical construction, it knows no limit. Only our imagination can stump us. These manipulations of mappings between all kinds of sets are just a series of steps one must follow like a recipe to produce a delicious cake.

 Here's a mapping that has characterized our present civilization:

Alphabet/number $\rightarrow$ base two numbers (0,1)

$\rightarrow$ switches (on/off)

And that advancement led us to build computers and the internet.

Here's something to know about imaginary numbers: anything that repeats itself – a circle, a wave – can best be described by an imaginary number. This illustrates that any of our number systems can have a specific geometric interpretation.

As we have indicated above, take for instance the irrational numbers: they arise because our minds invented the concept of area. When in doubt, ask yourself: does the universe care that we, humans, define an area of a square of size "b" as – "b times b", or "b^2"? Absolutely not, the concept of an area is our invention. Well, when we draw a square if one unit, the diagonal of that square is inevitably root 2, or written as $\sqrt{2}$ or $2^{\frac{1}{2}}$. Irrational numbers pop out of that geometric situation – this is a discovery that logically came out of our invention of an area, which is an abstract mental construct that has come from our creative mind.

On an imaginary axis (Fig. 6.1 below) $e^{i\alpha}$ describes a circle. Waves are described by functions of the type, $e^{i(kx-wt)}$, complimentary gift from Fourier who established the important idea, which roughly states: given any frequency of an oscillatory motion, as crazy as that motion can be, one can represent that as a series of sine and cosine functions, which is what $e^{i(kx-wt)}$ is. Those sine/cosine functions are perfect for ideal waves, as they have well-defined wavelengths and frequencies. I doubt very much they exist in nature – likewise for the perfect circle or the perfect square.

As we will explore, time is linked to the inner moving parts (IMPs) of a clock, for which motion is regularly repeated so its IMPs travel the same distance for each cycle. And as it was mentioned above, such regular motion is more than adequate to be described by complex numbers.

Another mental construct are coordinates: you look around, and they don't exist except as a human construct. Similarly, time doesn't exist, except as a human construct. And therefore a space-time construct is more involved: we need a ruler to establish the x,y,z coordinates for the spatial axes, but then we need a clock for the temporal axis. However here's a secret: in reality, the so-called "time" axis in Special Relativity (SR) is "ct", where c is the speed of light, making that "temporal" axis in actuality a spatial axis as "ct" is the distance travelled by light. But we will say more about this in the next chapter.

1.4 Motion and Time

Earliest civilizations date about 10,000 years while the earliest signs of the human species go back as far as 200,000 years ago. We can safely say that humans invented the concept of time way before the first signs of civilization. They defined time not knowing that they were observing some kind of motion. A day was really the full rotation of the earth on its axis, and a year, the full rotation of the earth around the sun. Mathematically, what our ancestors did can be represented as a mapping:

Observation of the seasons, changing daylight → time

Time and motion are two sides of a coin. It should not be a mystery that the rotation of the needles on a clock mimics the rotation of the earth on its own axis.

Rotations of the needles (clock/time) →

rotations of the Earth (motion)

Why we label the first (rotations of the needles) as time is just a convenient, cultural and practical thing.

Now space is easy to deal with – draw two points, use a stick and measure the distance – according to some standard stick kept in a vault somewhere in Paris. But measuring motion is an entirely different matter. What we do, and have done since Galileo, is to compare the motion of a given object with the motion of the internal moving parts (IMPs) of a clock – those moving parts are deliberately set to give regular beats, hence regular motion, that is, the "internal moving parts" (IMPs) travel the same distance after each tick the clock gives out. The need to standardized motion, which was an urgent preoccupation for Galileo, is the result of an observation that there are different kinds of motions – uniform motion versus non-uniform. It is the latter that prompts us to ask: "what causes a non-uniform motion (acceleration)?" To which the notion of force comes into play. Later on in the development of physics, the notion of this cause-and-effect would be rescheduled as a transfer of energy.

Now we want to take a closer look at what takes place when we decide to actually measure motion. For instance, if the motion of the given object under observation happens to give a constant ratio of distance travelled by the object with respect to the clock's IMPs, we know automatically that the observed object is in uniform motion, meaning it has a constant speed. This is the only way to measure – that is, quantifying – motion. In fact, here is what we have been doing in calculating this ratio:

$$\text{Ave. speed} = \frac{\text{Total distance traveled by the object}}{\text{Total distance traveled by the IMP of the clock}}$$

1.1

In Special Relativity (SR), we frequently use c=1, giving the units of time and length the same dimension. SR is just a rediscovery of what had been done unknowingly in the study of motion.

We can symbolize this process of observing the motion of a car, and in trying to figure out its speed (motion) by means of a table:

1	2	3	4
Clock ticks	Distance of clock's internal moving parts (IMP) in a tick	Distance of moving car in a tick	Ave. speed
t_1	d_{IMP}	d_1	d_1 / d_{IMP}
t_2	d_{IMP}	d_2	$(d_1+d_2)/ 2d_{IMP}$
.			
.			
.			

Fig. 1.2

From this we can construct the following diagrams for a body in motion:

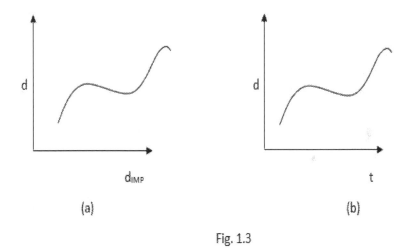

(a)   (b)

Fig. 1.3

The diagram in Fig. 1.3b is the most familiar as this is the one used in textbooks to define the speed of a body in motion. But it is a faithful reproduction of Fig. 1.3a when you take into consideration the following mapping,

$$d_{IMP} \rightarrow t \hspace{4cm} 1.2$$

It turns out that diagram in Fig. 1.3a is closer to reality – that is, if one opens the clock, one sees the moving parts of a clock. On the other hand, time is just a mental construct born out of that particular motion.

Note: if the clock gives regular beats, it simply means that its IMPs have travelled the same distance over and over – Column 2, d_{IMP}. If the ratio is the same throughout – Column 4 – then we know that the moving object is undergoing uniform motion   its velocity is constant, that is, there is no acceleration, and we can deduce from Newton's laws of motion that the net force acting on it is zero.

At every tick of the clock, the IMPs will cover the same distance over and over. This is a standardized distance locked into the device. So every motion that is understudied, its distance is measured against that standardized distance. Just like a stick kept in a vault in Paris is the standardized measure of distance, the second (tick) is the standardized measure of motion. A clock is simply a device that provides a standardized motion against which all other motions are compared.

In SR, time slows down because clocks slow down. There is no alternative otherwise we would get an object traveling faster than light, a phenomenon never observed so far. We will tackle the implications of these considerations on SR in the next chapter.

Now, Galileo was concerned with precise clocks because his main focus was to identify different types of motion: in particular, uniform motion (constant speed) and uniform acceleration which he studied by rolling balls over an inclined plane, and the acceleration due to gravity. In effect, what he did was to change the above mapping:

Motion (real world) → time (mental construct)

Newton took us on the wrong path by declaring that time was absolute, when in fact, motion is relative, and therefore time can only be relative in kind. But it took nearly 300 years before Einstein corrected that error.

Now in most standard textbook, speed was defined as distance travelled over a time interval. It became customary to think of space and time as fundamental, and speed (motion) as a derived notion. A contemporary of Einstein, Minkowski, re-interpreted his work as a 4-D world in which space-time can play a dynamic role. And that cemented the prevailing thought that space-time was a fundamental concept, maintaining the status of motion as a derived concept.

But what is a clock if nothing but an apparatus made up of a specific type of motion? The fact is when one uses a measure of time (second, day, year, etc.) this is just a translation of a particular motion. "I'll see you in one hour" means "I'll see you when the earth will have completed 1/24th of its rotation around its own axis." Or "I'll see you in one year" is a translation of "I'll see you when the earth will have completed one full rotation around the sun". What about my grey hair at age 70? Isn't that a testimony of the passage of time? This grey hair is a testimony that I've travelled 70 times around the sun. While that travelling took place, stuff happened, one of which is my hair turning grey.

What about the notion that motion is a relative quantity and so, how can it be more fundamental than time?

True that motion is relative with respect to some reference point. An object can be in motion with respect to one point, and at rest with another point. But we must take into consideration the whole picture. For instance you and I can be at rest, but our hearts, lungs and the blood circulating in our body are not. The atoms in our body are in motion, and so are the electrons in those atoms. Now suppose everything that constitute your body and mine – cells, atoms, electrons, etc. – are all at rest. Extend that to every object in the universe. In such a universe where there is no motion, the concept of time is no longer possible as "one moment in time" cannot be distinguished from the next "moment in time".

1.5 Causality

The concept of causality can be defined as the following: A causes B, causes C, causes D... and so on. Event B is called the effect of cause A, and so on. Another way to express that is: A precedes B, which precedes C, which precedes D... and so on. We can illustrate this as a series of points, dispensing of the notion of time for the moment.

Fig. 1.4

So when we say event A precedes event B, in reality, event A represents a gazillions of events that took place simultaneously. Also there are a gazillion number of events that took place between A and B, which we can represent as $A_1...A_2...$, each of those subsets representing a gazillion of events. In the set called A, a subset of it can be the direct cause of a subset of events in the set A_1. Some in A are totally unrelated to any events in A_1. Similarly, in subset A_2, there is a subset of events which are the effects of some events in subset of A_1, but also the effects of some events of set A, which were not in any way related to events in subset A_1. We can see as we proceed from A to B to C, and so on, that causality as initially described in our definition involves more than we can chew. We can safely say that Fig. 1.4 is an oversimplification of a concept that is already a very complicated matter.

1.6 Past, Present, and Future

The past and future can best be comprehended by examining the trajectory of an object.

Consider a moving object, and since this illustration is static and not a video, pretend that it is right now at position 2, frozen in that position, also labelled "present" in Fig. 1.5, and positions 1 and 3 are labels for past and future positions respectively.

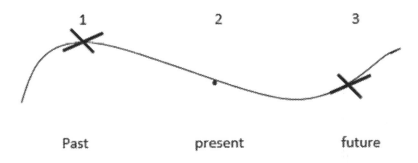

Past   present   future

Fig. 1.5

If you're asking, "Is time a one-way street?" you are asking the wrong question. Grabbing the object presently at position 2 and returning it to position 1 would not be a case for traveling back into the past – we're just altering the motion of that object to a different trajectory. The past doesn't exist, except as images of that motion either in our memory or on some form of media. What existed at position 1 was a motion that has continued into the present (position 2). The future (position 3) represents a motion of the object along its trajectory. It hasn't happened yet and therefore like the past, the future does not exist in the present in the sense that its content hasn't come about. In Fig. 1.5, position 3 represents a possible outcome. That such points as 1 and 3 are designated as past and future respectively does not endow these positions with existence. They are just a convenient way to talk about motion. Time is just a label for arbitrary points on a trajectory. Note that this is again what we fundamentally observe in the world: "matter moving through space". Time is a concept that helps us to visualize that fundamental observation.

1.7 Time and Temperature

In the linking of Quantum Field Theory (QFT) and Statistical Field Theory, there is an interesting mapping[1]:

$$\frac{it}{\hbar} \rightarrow \frac{1}{k_B T}$$

Where k_B is Boltzmann's constant, $\hbar$ is Planck's reduced constant, and i is the complex number.

At first sight, this might be odd, to say the least. But considering that time t is a measure of motion; and temperature T is a measure of the kinetic energy, which is energy of motion, then underlying this odd mapping is just, "something about motion" → "something about motion".

1.8 Conclusion

This new proposal that time is a derived concept demands a rethinking of an idea that has stretched back to the Ancients. And so this is difficult to absorb as we were trained to think of time differently. Many live by the calendar. However we must keep in mind that a universe without motion is a universe without time. Motion is the fundamental thing that we observe. Without motion, the idea of time would be inconceivable. And so it must be acknowledged that motion is a fundamental concept. And time is a derived concept, which is nevertheless indispensable in making the measurement of motion possible. With its IMPs, a clock and its regulated time measured in seconds is basically standardized motion against which all other motions are measured. The fact is that in just about every situation the word "time" can be replaced by an object undergoing a specific motion, and there would be no loss in meaning. The bottom line is that we measure the motion of a body with a standardized motion, which is what a clock offers.

There is a crack in everything, that's how the light gets in.

- Leonard Cohen

Chapter 2

New Insights into Special Relativity

(Did Einstein mean to say something different?)

In view of what was discussed in the previous chapter about the nature of time, we must now revisit the concept of space-time, also known as a Minkowski diagram or coordinate system, the concept being ubiquitous in both Special Relativity (SR) and General Relativity (GR). As you might guess, the Minkowski space-time concept needs a thorough examination. But before we tackle that task, we must look at the Lorentz transformation laws, which were developed before the Minkowski coordinate system made its appearance.

It has been said that after Einstein had published his seminal papers on SR in 1905, Minkowski had rewritten his theory in 4-D such that Einstein himself had mildly complained that he didn't recognize his own theory. This has since been corrected with a (1,3) designation instead of 4-D in recognition that time is different – at least it transforms differently than space in the mathematical sense.

Let's put some clarity into this seemingly messy concoction.

We start with the fundamental assumptions of SR known as the two postulates.

2.1 The Postulates of Special Relativity

Special Relativity is based on two basic postulates:

Postulate 1: The laws of physics are the same in all inertial frames.

Postulate 2: The speed of light is invariant in every inertial frame.

Note 1: The main point for postulate 1 is about observers in different frames. Two observers can choose whatever coordinate system – both chooses the same (Cartesian) or different (one, Cartesian; the other, polar) – but what we must keep in mind is that we are dealing with different frames, regardless of the choice of a coordinate system.

Note 2: "Invariant" means that a measurement in one inertial frame will be the same in a different inertial frame, while "constant" means that it doesn't change in any frame. To see this more clearly, consider the 3-D momentum, p_i with i = 1,2,3. In an elastic collision, the total momentum before collision is equal to the total momentum after collision. The total momentum is said to be a constant. In a different frame, a second observer will also observe that the total momentum before collision is equal to the total momentum after collision. But this constant will be different than the constant observed by the first observer. The 3-D momentum is not a Lorentz invariant. However, the space-time interval is a Lorentz invariant as we shall see in this chapter, but it is not a constant quantity. In the case of the speed of light, it is both a constant and a Lorentz invariant.

At the heart of these two postulates are the Lorentz transformation laws to which we now turn our attention. Since light moves in a spherical wave front, it can be described by observer O as a sphere in that inertial frame,

$$c^2t^2 = x^2 + y^2 + z^2 \qquad 2.1$$

We can rewrite this as,

$$c^2t^2 - x^2 - y^2 - z^2 = 0 \qquad 2.2$$

A second observer O' in a different inertial frame will also write,

$$c^2t'^2 - x'^2 - y'^2 - z'^2 = 0 \qquad 2.3$$

Since these two equation are equal to zero, they are equal to each other. We have,

$$c^2t^2 - x^2 - y^2 - z^2 = c^2t'^2 - x'^2 - y'^2 - z'^2 \qquad 2.4$$

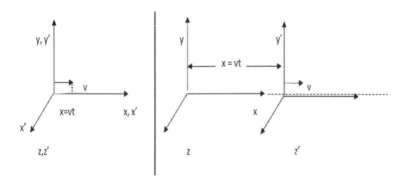

Fig. 2.1a   Fig. 2.1b

We can always arrange our two frames such that initially we have y' = y and z' = z, (Fig 2.1a).In Fig. 2.1b, the two frames are at a certain distance away from each other as the primed frame has moved a distance x, moving at a speed v going towards the right with respect to the unprimed frame.

In fig. 2.1a, equation 2.4 then becomes,

$$c^2t'^2 - x'^2 = c^2t^2 - x^2 \qquad 2.5$$

Another assumption is that the transformation laws that relate the two frames O and O' are linear. Therefore we can state that,

$$x' = Ax + Bct \qquad 2.6$$

$$ct' = Cx + Dct \qquad 2.7$$

Where A, B, C, and D are to be determined. Those two equations can be put as a single matrix equation such as,

$$\begin{pmatrix} x' \\ ct' \end{pmatrix} = \Lambda \begin{pmatrix} x \\ ct \end{pmatrix} \qquad 2.8$$

Where $$\Lambda = \begin{pmatrix} A & B \\ C & D \end{pmatrix} \qquad 2.9$$

Substituting equations 2.6 and 2.7 into the LHS of equation 2.5, we get

$$c^2t'^2 - x'^2 = (Cx + Dct)^2 - (Ax + Bct)^2$$

Expanding and collecting similar terms we have,

$$c^2t'^2 - x'^2 = (D^2 - B^2)c^2t^2 - (A^2 - C^2)x^2 + 2(DC - AB)xct$$

In order to satisfy the RHS of equation 2.5, we need

$$D^2 - B^2 = 1$$

$$A^2 - C^2 = 1$$

$$DC = AB$$

This can be satisfied by using the following identity,

$$cosh^2\theta - sinh^2\theta = 1 \qquad\qquad 2.10$$

And then we set,

$$A = D = cosh\theta, \; B = C = -sinh\theta \qquad 2.11$$

Other settings would depict motions different than the one in Fig. 2.1.

We can now solve for the angle θ. We note that when the two origins are coincident (Fig. 2.1a), x' = 0 and x = vt. Substitute that with equations 2.11 into 2.6, we get,

$$x' = 0 = (cosh\theta)x - (sinh\theta)ct$$

$$= (cosh\theta)vt - (sinh\theta)ct$$

$$\tanh\theta = \frac{v}{c} \qquad\qquad 2.12$$

Using the identity,

$$\cosh\theta = \frac{1}{\sqrt{1-tanh^2\theta}} \qquad\qquad 2.13$$

We get

$$\cosh\theta = \frac{1}{\sqrt{1-\frac{v^2}{c^2}}} \equiv \gamma \; ; \sinh\theta = v/c \qquad 2.14$$

Note that for massless particle, v = c and γ = 1; for massive particle, v < c and γ > 1.

We can now express the Lorentz transformation equations (2.6 and 2.7) in the familiar form:

$$x' = \gamma(x - vt);$$

$$ct' = \gamma\left(ct - \left(\frac{v}{c}\right)x\right);$$

$$y' = y;$$

$$z' = z \qquad\qquad 2.15$$

Before we proceed, we must carefully examine equation 2.1, reproduced below as,

$$(c^2t^2)_{light} = (x^2 + y^2 + z^2)_{light} \qquad 2.16$$

We must reiterate that both sides of the equation referred to the distance squared traveled by light in the unprimed frame. This was done in a 3-D Cartesian coordinate system (also being a flat space) with time as a parameter.

Now an important concept is the space-time interval based on equation 2.2, which is defined as,

$$ds^2 \equiv -c^2dt^2 + dx_i\, dx^i \qquad 2.17$$

Where we now use differentials since we are going to be dealing with distances, rather than positions. The Einstein summation is applied, and we set $x = x^1, y = x^2, z = x^3$.

Using what is called the 4-vector formalism, we define a 4-vector $x^\mu = (x^0, x^i)$ where Latin index $i = 1,2,3$; and Greek index $\mu = 0,1,2,3$, and $ct \equiv x^0$.

The metric tensor is a 4×4 matrix,

$$\eta_{\mu\nu} = \begin{pmatrix} -1 & 0 & 0 & 0 \\ 0 & 1 & 0 & 0 \\ 0 & 0 & 1 & 0 \\ 0 & 0 & 0 & 1 \end{pmatrix} \qquad 2.18$$

With these tools in hand we can rewrite equation 2.17 as,

$$ds^2 = \eta_{\mu\nu} \, dx^\mu \, dx^\nu \qquad 2.19$$

Note 1 - the thumb rule to remember is: no index for a scalar (s); one index for vectors (x_μ); and two indices or more for tensors ($\eta_{\mu\nu}$). The "2" in ds^2 is an exponent, not an index.

Note 2 – the metric tensor is used to lower or raise indices:
$$A_\mu = \eta_{\mu\nu} \, A^\nu$$

Note 3 – the norm of a vector is defined as

$$|A| = (\eta_{\mu\nu} \, A^\mu \, A^\nu)^{½}$$

2.2 The Invariance of the Space-Time Interval

The invariance of the interval is so important that it crystalizes everything about SR. No physical theory of the real world can survive without it.

We rewrite equation 2.8 for differentials with the indices in the right places.

$$dx'^\mu = \Lambda^\mu_\nu \, dx^\nu \qquad 2.20$$

Our assumption that the transformation laws relating the two frames O and O' to be linear demands that the Λ's have an inverse expressed by this restriction:

$$\Lambda^\mu_\gamma \Lambda^\nu_\delta \eta_{\mu\nu} \;=\; \eta_{\gamma\delta} \qquad\qquad 2.21$$

In the primed frame, observer O' writes down her own equation for the interval as,

$$ds'^2 \;=\; \eta_{\mu\nu}\, dx'^\mu dx'^\nu \qquad\qquad 2.22$$

Note that the metric tensor, composed of 0's and 1's, is invariant in this particular case. In GR, $\eta_{\mu\nu} \to g_{\mu\nu}$, and so this will no longer be the case.

Substituting equation 2.20 into 2.22, we get

$$ds'^2 \;=\; \eta_{\mu\nu}\, \Lambda^\mu_\gamma\, dx^\gamma \Lambda^\nu_\delta\, dx^\delta \qquad\qquad 2.23$$

Using the above restriction (equation 2.21), we obtain

$$ds'^2 \;=\; \eta_{\gamma\delta}\, dx^\gamma dx^\delta \qquad\qquad 2.24$$

Now the indices in the Einstein summation are dummy indices, the RHS is just equation 2.19. Therefore,

$$ds'^2 \;=\; ds^2 \qquad\qquad 2.25$$

We now see the power of the 4-vector formalism. In that mathematical framework, the interval ds is an invariant: any observer in any initial frame will measure the same interval.

Some general comments:

(1) More specifically, what we have is a space-time coordinate system that is mapped into a manifold, which we represent as:

Space-time → manifold

Now, mathematically, a manifold is a surface with any shape or form that fulfills certain mathematical conditions, but there are some that are more useful than others. All observations indicate that space is flat, and not curved, a popular misconception. It's the combination of space and time that is best represented mathematically on a curved manifold. Similarly, quantum states are mapped as vectors (rays) in a Hilbert space, and not as vectors in the real world. Manifolds, Hilbert space – these are abstract mental constructs.

(2) Another look at this would be such mappings as:

Real world → manifolds

(3-d) space manifold → (4-D) space-time manifold

GR: real world → space-time manifold (STM)

QM: real world → internal space manifold (ISM)

Both STM and ISM have similar mathematical structure but they are different as apples and oranges. For instance, STM depends on the metric tensor $g_{\mu\nu}$ which is sensitive to coordinate transformations, and hence the choice of a coordinate system.

(3) Some people will quibble about why the insistence on stating that a 3-D space of our world is real while a (4-D) manifold is an abstract construct. Now a 3-D space manifold is closer to the real world for several reasons. The first reason is if you lay out three independent sticks to form a volume, a fourth one can only be expressed as a combination of the first three. This is not a mathematical theorem but an observation, that's part of our real world.

There is no escape from that reality. The second reason is what we've explained in chapter one: time is not a fundamental feature of the universe. It is the midwife that allows us to measure motion. And thirdly, the time in SR is really "ct", a distance, which we will give more details below.

Now the 4-vector formalism is convenient for our needs – it makes the equation more compact, more esthetically pleasing, and easier to manipulate algebraically. But more importantly, the value of working in a 4-D or (1,3) space-time manifold is that the interval ds is an invariant. We construct a 4-D space-time mathematical model not because the universe is 4-D but because if we find one true equation in SR using the 4-vector formalism then we know it is true in any other frame. This property is referred as Lorentz invariance which is basically an insurance policy to make sure that any equation written in any inertial frame will have the speed of light as a constant.

(4) We need to also point out that for a massive particle, moving along a straight line in the x-direction, we have,

$$ds^2 = -c^2dt^2 + dx^2 < 0 \qquad 2.26$$

That's true because no massive particle can travel faster than light, that is, $dx < cdt$.

2.3 The Minkowski Space-Time Diagram

Note: I will call it the Minkowski diagram instead of the Minkowski coordinate system for reasons I will elaborate. The conventional Minkowski diagram is one in which time is no longer a parameter as in Fig. 2.1, instead it is presented as a coordinate which to all intents and purposes seems to reside outside of space?!? (Fig. 2.2)

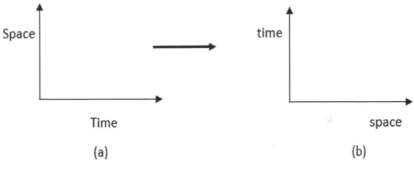

(a)   (b)

Fig. 2.2

Note that in Fig.2.2a is a graph which is what is most familiar in elementary physics. It is on this graph that velocity is defined as the ratio of distance/time. In SR, that graph is rotated to yield Fig. 2 b, and is called a coordinate system. There are issues with that. Many people have convinced themselves that they understand this diagram. But it's a lot trickier than what it looks like. To say the least: a Minkowski diagram is a very unusual hybrid. Moreover, it cannot exist in the real world: 1) as we have said in the previous chapter, time is standardized motion, expressively conceived to measure the motion of another object which can only take place in a 3-D space; 2) nothing can exist outside of space - everything is "matter moving through space".

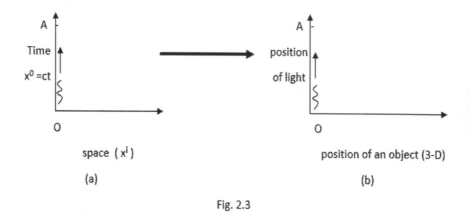

Fig. 2.3

So why use a Minkowski coordinate system? Its usefulness resides on the fact that equation 2.19 is inadvertently linked to the Lagrangian function, although this was not the original intention when the concept of space-time came about. We will go over what the Lagrangian is in later sections. For now, needless to say, the hybrid Minkowski space-time took a life of its own.

So let's take a very careful look on how one should read the Minkowski diagram (Fig. 2.3).

A burst of light leaves the origin O and travels through space to a point A along the vertical axis (the word "time" replaced by the position of light). However, the position of an object along the space axis refers to an object placed in a 3-D Cartesian coordinate system, which can be different than light.

There is nothing wrong in making the following mapping for an observer O in some given frame,

$$ct \quad \rightarrow \quad time \quad \rightarrow t$$

Since according to postulate 2, the speed of light in a different frame will be the same (c = c'), an observer O' in a different inertial frame can also make a similar mapping,

$$c't' \quad \rightarrow \quad time \rightarrow t'$$

In some textbook, the speed of light is set such that c =1, which is also fine as dimensional units are arbitrary and one can always design a system of units in which c = 1.

But most importantly, not to be forgotten is that $x^0 = ct$ measures the distance that light travels in a given time t in that frame. Moreover if the light is emitted from the origin than "ct" is also the position of light at time t in that frame. Time is NOT flowing from position O to position A instead we have a burst of light moving through space from O to A (Everything is "matter moving through space"). This means that the axis, $x_0 = ct$, is a vertical spatial axis, no doubt different that the horizontal spatial axis, but nonetheless spatial in kind.

The other objection we might have is why place the ct-axis orthogonal to the space axis. Perhaps for convenience. But we must be careful not to interpret this as time existing independently outside of space. If we superimpose the primed frame, which is moving away from the unprimed frame at a velocity v, we get,

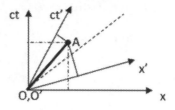

Fig. 2.4

We see that the ct'-x' coordinate system is no longer orthogonal viewed from observer O. Note that both coordinate systems are symmetric about the same line x=t, x'=t', that's because this is the case of the interval being equal to zero in both observers' frames, which is the trajectory of a massless particle. But most importantly the interval OA and O'A, which have different components in their respective frame, is nevertheless invariant. Both observers will measure the same interval $(OA)^2 = (O'A)^2$.

Also the Minkowski coordinate system leads to some peculiar geometry. For instance, in a Euclidean space, the Pythagoras equation prevails.

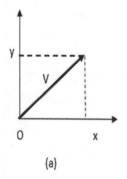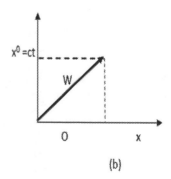

(a)   (b)

Fig. 2.5

In Fig. 2.5a, we have for the vector V,

$$V^2 = x^2 + y^2 \qquad 2.27$$

While on a Minkowski coordinate system, Fig. 2.5b, we have (note 3, section 2.1),

$$W^2 = -(ct)^2 + x^2 \qquad 2.28$$

Now some people have claimed that this is an indication that space-time is non-Euclidean, and by extension, the universe is also non-Euclidean. Consider that the Minkowski is an unconventional hybrid, I'm not willing to go that route. The Minkowski coordinate system describes two separate objects: a burst of light on one axis, and another separate object in the spatial axis. It is not a coordinate system in the same sense as a Cartesian system, and therefore such a claim - the universe being non-Euclidean - is misleading. Nevertheless the Minkowski diagram has its merits as a mathematical tool. And we should treat it as no more and no less than that: a graph which is a useful mental construct.

To clarify this point, recall that in chapter 1, we've indicated that math is primarily the mapping between sets, and there is an infinite soup of these mappings that can be created from our imagination, and that mathematicians have the task of sorting these out. So here's a diagram of two sets describing a population versus the age of the individuals composing that population.

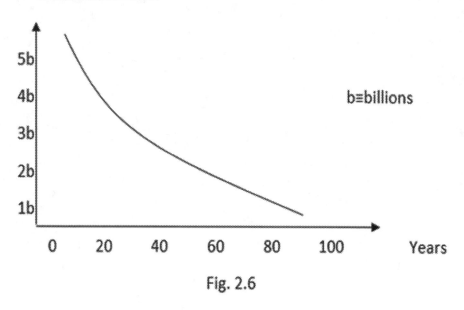

Fig. 2.6

We get valuable information about a population decreasing in age. We can even make the prediction that everyone will eventually die. But no one would claim that it describes the universe in term of the physical equations that govern the universe. Here's a more appropriate diagram that is closer to the physical equations describing the real world:

Case B: The Coordinate System

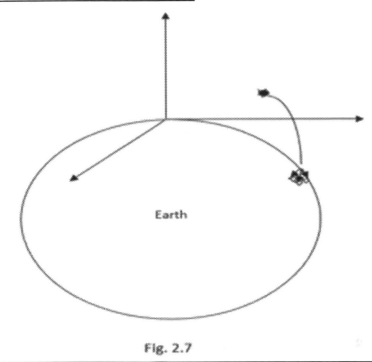

Fig. 2.7

In Fig. 2.7, we have an observer at the pole observing the trajectory of a shooting star crashing on the planet. It's a Cartesian coordinate, and time (standardized motion) is a parameter. This is a coordinate system in which we can express the laws of physics.

The Minkowski coordinate system is a hybrid that falls between Case Λ and Case B, valuable but not a reliable description of the real world in the strict sense of Case B. As to its coordinates, their interpretation is not what has been claimed: the "ct" axis is a spatial coordinate of light moving a distance ct, even though one can map this distance to a time-axis, while the x's make up a spatial coordinate for some other object.

To put it in another way:

(1) It is a plot between two sets (Case A: graph).

(2) It is a coordinate system with ct → t (Case B). However in a coordinate system, a single point represent a single object on both axis, vertical and horizontal. But in a Minkowski diagram, a single point refers to the position of light on one axis and another object on the other axis that can be different than light.

Case C: The Complex Plane

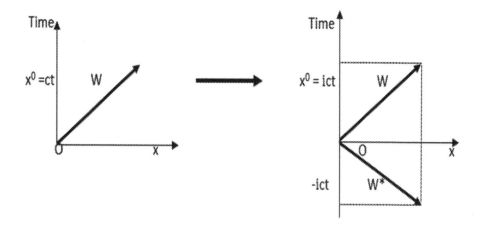

Fig. 2.8

The complex plane is known as an Argand diagram. The y-axis becomes y → iy, where $i = \sqrt{-1}$.

In the Minkowski diagram (fig. 2.8), we can map $x^0 = ct \rightarrow x^0 = ict$. Applying Pythagoras theorem we get back equation 2.28, $W^2 = (ict)^2 + x^2 = -(ct)^2 + x^2$.

The Minkowski diagram is akin to an Argand diagram but not completely. The norm of the vector **W** on a Minkowski diagram is $(W^2)^{\frac{1}{2}} = ((-ct)^2 + x^2)^{\frac{1}{2}} \equiv (\eta_{\mu\nu} x^\mu x^\nu)^{\frac{1}{2}}$, while on a complex plane the norm is defined as $(WW^*)^{\frac{1}{2}}$, where W* is the complex conjugate, $x - ict$. The norms are defined differently, that the answer turns out to be the same is a coincidence. The complex plane can be expressed in Cartesian coordinates (x,y) or polar coordinates (r, θ), but nevertheless it is not a coordinate system, it is a graph of complex functions.

The Minkowski diagram is problematic. When using it, one must be vigilant and careful as to the claims that can be made. The Minkowski diagram does have its merit – it illustrates quite nicely that the space-time interval is Lorentz invariant - but one shouldn't be blinded because the math appears to be so appealing.

2.4 Time Dilation, Twin Paradox, Length Contraction

Even though time dilation mostly grabbed the main headlines of Einstein's revolutionary ideas after 1905, underlying that concept is length contraction. As we have argued, a clock delivers standardized motion through the distance its IMP traverses as it gives out its familiar, regular ticking sound. It goes without saying then that length contraction is the underpinning of time dilation.

To illustrate these two phenomena, consider again two observers O and O', Fig. 2.1. In the primed frame, the clock is at rest with respect to O', and necessarily moving with respect to O.

Recall equation 2.5 reproduced below with some modifications:

$$c^2 \Delta\tau'^2 - \Delta x'^2 = c^2 \Delta t^2 - \Delta x^2 \qquad 2.29$$

We are using the Δ symbol to indicate that we are looking at intervals of time and position; and the symbol τ for the proper time, which is defined for a clock at rest. From the point of view of O', the clock is at rest, and so we have $\Delta x'^2 = 0$. Therefore,

$$c^2 \Delta\tau'^2 = c^2 \Delta t^2 - \Delta x^2 \qquad 2.30$$

Divide both sides by c^2, and factor out Δt^2. We get,

$$\Delta\tau'^2 = \Delta t^2 \left(1 - \frac{\Delta x^2}{c^2 \Delta t^2}\right) \qquad 2.31$$

Using $v = \frac{\Delta x}{\Delta t}$ and taking the square-root on both sides, the result is

$$\Delta\tau' = \Delta t \left(1 - \frac{v^2}{c^2}\right)^{\frac{1}{2}} \qquad 2.32$$

From equation 2.14, we have

$$\Delta\tau' = \Delta t / \gamma \qquad 2.33$$

With γ > 1, the proper time $\Delta\tau' < \Delta t$, we say that moving clocks slow down.

In the twin paradox, the frequent question asked is: how do we determine which clock registers the proper time, since a clock can be at rest with respect to one observer but is in motion with respect to another observer.

Consider Alice is in a rocket ship about to undergo a trip to the nearest star to our sun, say planet Proxima Centauri in the Alpha Centauri star system – about 4.22 light-years. Her twin brother Bob stays back home on planet earth. In this case, the clock in Alice's ship will register the proper time. Moving clocks slow down. How do we come to that conclusion? Alice will register two events with one clock: her departure from earth and her arrival on Proxima Centauri. Her clock is at rest with respect to Alice but not with respect to the two events – it has to move from event one (departure) and event two (arrival). On the other hand, Bob will need two clocks to measure the time taken by his sister of these two events: one clock on planet earth, and another one on Proxima Centauri – he needs his good friend Cortney already on Proxima Centauri to record Alice's arrival. So we can see that Bob's clock is NOT the moving clock as it stays back home, registering just one of those two events. Therefore it is Alice who will look younger in age than her twin brother Bob as her clock and everything in her rocket ship will experience this time dilation. Note that the twin paradox is not about traveling to the future but it's all about aging. Bob will age faster than Alice.

This phenomenon has been confirmed in the decay of muons. Comparing the half-life decay between muons producing in the lab with those arriving at very high speed from outer space was a triumph of SR.

In the case of length contraction, we use a car moving at a speed v. Observer O will measure the length of that car with one clock by registering the time as both front and end of that car will pass in front of him.

$$L = v\,(t_2 - t_1) = v\,\Delta t \qquad\qquad 2.34$$

Observer O′ is sitting in the car, but in this case, he needs two clocks, one in the front and another in the back of the car, as his car passes through right in front of observer O. He will also measure,

$$L' = v\,(t'_2 - t'_1) = v\,\Delta t' \qquad\qquad 2.35$$

We can see that it is observer O who registers the proper time (he uses one clock),

Hence equation 2.34 becomes,

$$L = v\,(t_2 - t_1) = v\,\Delta t'/\gamma \qquad\qquad 2.36$$

Divide equation 2.36 by 2.35 yields the result for length contraction,

$$L = L'/\gamma \qquad\qquad 2.37$$

With γ > 1, L < L′. The moving car (the ruler) is contracted, hence moving rulers contract.

As we have argued in chapter 1, the speed of an object is, (equations 1.1 and 1.2 reproduced below)

$$\text{Ave. speed} \;=\; \frac{\text{Total distance traveled by the object}}{\text{Total distance traveled by the IMP of the clock}}$$

$$d_{IMP} \;\rightarrow\; t$$

From this we see that time dilation is length contraction in disguise. In the twin paradox, the dual action of length contraction and time dilation is at work. According to Bob,

Alice will have a slower clock but will also measure her distance with a shrinking ruler. Therefore both Alice and Bob will agree that the rocket ship (the frame) in which Alice is traveling moves at the speed v. This is an important result. As you might have guessed the interval is Lorentz invariant, but not the speed. However, with the cancellation of the γ factor in the dual action, the speed of the frame is invariant. From Fig. 2.1, we chose the observer O at rest, and observer O' moving to the right. That choice was arbitrary as we could have easily chosen O' at rest and O moving to the left. Had the case be that the speed of the frame was not invariant – O and O' would be measuring their relative speed differently - SR as a physical theory would collapse.

2.5 Action, Lagrangian, Equations of Motion

An object S is said to be invariant if a change in S vanishes. That is, mathematically, if S is invariant then δS = 0.

The Lagrangian is defined as the difference between the kinetic energy and the potential energy of a particle.

$$L = T - V \qquad\qquad 2.38$$

Note that the Lagrangian is not necessarily conserved. What is conserved is the total energy, called the Hamiltonian, $H = T + V$.

Generally, the Lagrangian is a function of position and velocity, while the Hamiltonian is a function of position and momentum.

$$L \rightarrow L(q, \dot{q}) \qquad H \rightarrow H(q, p)$$

Where the dot indicates a derivative with respect to time.

The Action is defined in terms of the Lagrangian as:

$$S = \int L(q, \dot{q})\, dt \qquad 2.39$$

Note: lower case s for the space-time interval, upper case S for the action.

The principle of least action states that a particle will follow the path of least action. To find this path we minimize the variance of the action:

$$\delta S = 0 \qquad 2.40$$

Substituting equation 2.39 into the above:

$$\int \delta L(q, \dot{q})dt = 0 \qquad 2.41$$

Taking into consideration that we may have many particles, the LHS becomes,

$$\delta \int L(q, \dot{q})dt = \int \Sigma_i \left[\frac{\partial L}{\partial q_i} \delta q_i + \frac{\partial L}{\partial \dot{q}_i} \delta \dot{q}_i \right] dt$$

Integrating by parts the second term in the bracket, we get

$$\delta S = \int \Sigma_i \left[\frac{\partial L}{\partial q_i} - \frac{d}{dt} \left(\frac{\partial L}{\partial \dot{q}_i} \right) \right] \delta q_i\, dt \qquad 2.42$$

Since q_i is arbitrary, the above expression vanishes only if the term inside the bracket vanishes,

$$\frac{\partial L}{\partial q_i} - \frac{d}{dt} \left(\frac{\partial L}{\partial \dot{q}_i} \right) = 0 \qquad 2.43$$

These are called the Euler-Lagrange equations. What's so important about those equation is that they yield the equations of motion (eom).

EXAMPLE: take a single particle moving along the x direction with kinetic energy L = ½mẋ², into a potential V(x).

From equation 2.38, the Lagrangian is,

$$L = \tfrac{1}{2}m\dot{x}^2 - V(x) \qquad\qquad 2.44$$

From 2.43, we want to calculate the first term. We take the derivative of L with respect to x:

$$\frac{\partial L}{\partial x} = \frac{\partial}{\partial x}[\tfrac{1}{2}m\dot{x}^2 - V(x)] = -\frac{\partial}{\partial x}V(x) \qquad 2.45$$

To calculate the second term, we first take the derivative of L with respect to $\dot{x}$,

$$\frac{\partial L}{\partial \dot{x}} = \frac{\partial}{\partial \dot{x}}[\tfrac{1}{2}m\dot{x}^2 - V(x)] = m\dot{x} \qquad\qquad 2.46$$

Next we take the time derivative,

$$\frac{d}{dt}\left(\frac{\partial L}{\partial \dot{x}}\right) = m\ddot{x} \qquad\qquad 2.47$$

Putting everything together, we get

$$\frac{\partial}{\partial x}V(x) + m\ddot{x} = 0 \qquad\qquad 2.48$$

From classical physics, the force is

$$F = -\frac{\partial}{\partial x}V(x)$$

And $\ddot{x}$ *is the accelation*, we then have Newton's 2nd law of motion, F = ma.

The conjugate momentum is defined as,

$$p = \frac{\partial L}{\partial \dot{x}} \qquad\qquad 2.49$$

In the above example, from equation 2.46, p = $m\dot{x}$ = mv, which is the classical momentum.

Also, the Hamiltonian is defined through the Legendre transform,

$$H = p\dot{x} - L$$

2.50

In the above example,

$$H = (m\dot{x})\dot{x} - (\tfrac{1}{2}m\dot{x}^2 - V(x))$$

$$= \tfrac{1}{2}m\dot{x}^2 + V(x)$$

$$= T+V$$

2.51

And that is the total energy of the system.

2.6 The Space-Time Interval and the Lagrangian

In section 2.3, we asked: So why use a Minkowski coordinate system?

We made the point that its usefulness resides on the fact that equation 2.19, reproduced below, is inadvertently linked to the Lagrangian function.

$$ds^2 = \eta_{\mu\nu}\, dx^\mu\, dx^\nu$$

2.52

Where is that link coming from? We want to know how to extremize the path of a free particle (the principle of least action from the previous section) on a Minkowski diagram. This is fine as a Minkowski diagram is suited as a configuration space, on which the principle of least action is conventionally defined.

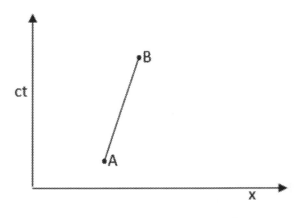

Fig. 2.9

We write for the proper time, using $d\tau^2 = -ds^2$ and equation 2.52:

$$\tau_{AB} = \int_A^B d\tau = \int_A^B (-\eta_{\mu\nu} \, dx^\mu \, dx^\nu)^{1/2} \qquad 2.53$$

We want to parametrize the line such that A = 0 and B =1, with a parameter σ.

$$\tau_{AB} = \int_0^1 (-\eta_{\mu\nu} \frac{dx^\mu}{d\sigma} \frac{dx^\nu}{d\sigma})^{1/2} \, d\sigma \qquad 2.54$$

We identify the above equation with the action (equation 2.39), reproduced below:

$$S = \int L(q, \dot{q}) \, dt \qquad 2.55$$

With the following mapping:

$$S \rightarrow \tau_{AB} \, ; \; L \rightarrow (-\eta_{\mu\nu} \frac{dx^\mu}{d\sigma} \frac{dx^\nu}{d\sigma})^{1/2} \, ; \; dt \rightarrow d\sigma \qquad 2.56$$

As we've seen in the previous section, setting δS = 0 yields the Euler-Lagrange equations (2.43) (reproduced below),

$$\frac{\partial L}{\partial x} - \frac{d}{d\sigma}\left(\frac{\partial L}{\partial \dot{x}}\right) = 0 \qquad\qquad 2.57$$

Here $x = x^\mu$ and $\dot{x} = dx^\mu/d\sigma$

The first term is,

$$\frac{\partial L}{\partial x} = \frac{\partial}{\partial x}(-\eta_{\mu\nu}\dot{x}^\mu\,\dot{x}^\nu)^{\frac{1}{2}} = 0$$

The second term yields,

$$\frac{d}{d\sigma}\left(\frac{\partial L}{\partial \dot{x}}\right) = \frac{d}{d\sigma}\left(\frac{\partial}{\partial \dot{x}}(-\eta_{\mu\nu}\dot{x}^\mu\,\dot{x}^\nu)^{\frac{1}{2}}\right)$$

$$= \frac{d}{d\sigma}\left(\frac{\frac{1}{2}\,(2)\dot{x}^\mu}{(-\eta_{\mu\nu}\dot{x}^\mu\,\dot{x}^\nu)^{\frac{1}{2}}}\right)$$

$$= \frac{d}{d\sigma}\left(\frac{1}{L}\frac{dx^\mu}{d\sigma}\right)$$

Putting the two terms together, we get

$$\frac{d}{d\sigma}\left(\frac{1}{L}\frac{dx^\mu}{d\sigma}\right) = 0 \qquad\qquad 2.58$$

However,

$$L = \left(-\eta_{\mu\nu}\frac{dx^\mu}{d\sigma}\frac{dx^\nu}{d\sigma}\right)^{\frac{1}{2}}$$

$$= \left(\frac{d\tau^2}{d\sigma^2}\right)^{\frac{1}{2}}$$

$$= \frac{d\tau}{d\sigma} \qquad\qquad 2.59$$

Substitute equation 2.59 into 2.58, and multiply throughout by $d\sigma/d\tau$, we get,

$$\frac{d\sigma}{d\tau}\frac{d}{d\sigma}\left(\frac{d\sigma}{d\tau}\frac{dx^\mu}{d\sigma}\right) = \frac{d^2x^\mu}{d\tau^2} = 0 \qquad\qquad 2.60$$

The solution is the equation of a straight line connecting A and B (eom). So the Minkowski diagram provides an open door to set the stage for QFT, which incorporated both QM and SR through the Lagrangian. Some of those features will be explored in later chapters.

2.7 (Optional) Einstein's Famous Equation $E = mc^2$

Two reasons in particular why I want to bring out the derivation of Einstein's famous equation:

(1) It is through this equation that nature allows the decay of matter. Similarly, it is also with this equation that we can accelerate particles to very high speed, and convert their kinetic energy into the creation of new particles. In a universe without the conversion of mass into energy, and vice versa, these two phenomena would be impossible.

(2) The method Einstein derived his equation by looking at events from two observers in two different initial frames was used extensively in this chapter. On a personal level that method led me to the discovery of a new law of kinematics, which will be the subject of the next chapter.

One more remark before we proceed: though Einstein was loudly praised for his creativity, his derivation of $E = mc^2$ is also a demonstration of impeccable logic.

Here's a quick rundown of that most famous equation.

Einstein knew from experiments previously done that a particle could decay and release gamma rays. He reasoned that when this happened, the particle would lose kinetic energy, and this could only be accounted by a loss of mass. So how did he come to that conclusion? As we have mentioned above, he analyzed the situation both in a rest frame and in a moving frame.

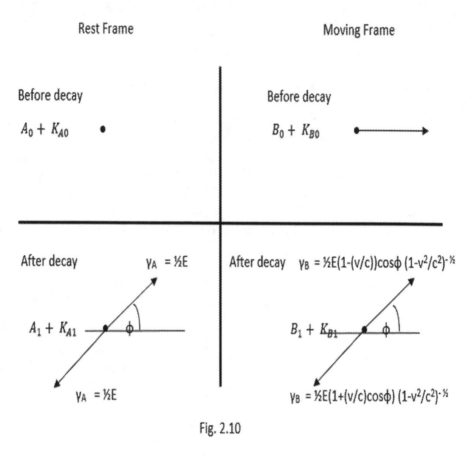

Rest Frame | Moving Frame

Before decay
$A_0 + K_{A0}$

Before decay
$B_0 + K_{B0}$

After decay $\gamma_A = \frac{1}{2}E$

$A_1 + K_{A1}$ ϕ

$\gamma_A = \frac{1}{2}E$

After decay $\gamma_B = \frac{1}{2}E(1-(v/c))\cos\phi \, (1-v^2/c^2)^{-\frac{1}{2}}$

$B_1 + K_{B1}$ ϕ

$\gamma_B = \frac{1}{2}E(1+(v/c)\cos\phi) \, (1-v^2/c^2)^{-\frac{1}{2}}$

Fig. 2.10

The symbol γ refers to the energy of a photon .The value of γ in Fig. 2.10 was already known by Einstein[2].

The particle has some internal property, which is not needed to be identified, and we labelled it as A in the rest frame, and B in the moving frame. The symbol K is for kinetic energy. We use the symbol "0" for the event taking place before the decay, and "1" for after the decay.

By the law of conservation of energy: The energy before decay = the energy after decay.

(1A) in the rest frame:

$$A_0 + K_{A0} = A_1 + K_{A1} + \tfrac{1}{2}E + \tfrac{1}{2}E$$
$$= A_1 + K_{A1} + E \qquad\qquad 2.61$$

(1B) in the moving frame:

$$B_0 + K_{B0} = B_1 + K_{B1}$$
$$+ \tfrac{1}{2} E \left(1 - \left(\frac{v}{c}\right) \cos\Phi\right) (1 - \frac{v^2}{c^2})^{-\frac{1}{2}}$$
$$+ \tfrac{1}{2} E \left(1 + \left(\frac{v}{c}\right) \cos\Phi\right) (1 - \frac{v^2}{c^2})^{-\frac{1}{2}}$$
$$= B_1 + K_{B1} + E(1 - \frac{v^2}{c^2})^{-\frac{1}{2}} \qquad\qquad 2.62$$

Now taking a look at the energy difference of the particle in each of the frame:

(2A) in the rest frame:

$$A_1 - A_0 = -(K_{A1} - K_{A0}) - E$$
$$= -\Delta K_A - E \qquad\qquad 2.63$$

(2B) in the moving frame:

$$B_1 - B_0 = -(K_{B1} - K_{B0}) - E(1 - \frac{v^2}{c^2})^{-\frac{1}{2}}$$
$$= -\Delta K_B - E(1 - \frac{v^2}{c^2})^{-\frac{1}{2}} \qquad\qquad 2.64$$

Whether the observer is at rest or moving with respect to the particle, the energy difference (in 2.63 and 2.64) should be the same, independent of the internal properties of the decaying particle.

$$- \Delta K_A \ - \ E = - \Delta K_B \ - \ E(1 - \tfrac{v^2}{c^2})^{-\frac{1}{2}} \qquad 2.65$$

Calculating the difference in kinetic energy:

$$\Delta K = \Delta K_A - \Delta K_B$$

$$= E\left(1 - \frac{v^2}{c^2}\right)^{-\frac{1}{2}} - E$$

$$= E\left[\left(1 - \frac{v^2}{c^2}\right)^{-\frac{1}{2}} - 1\right]$$

$$\approx E\left(1 + \tfrac{1}{2}\frac{v^2}{c^2} - 1\right)$$

$$= \tfrac{1}{2}E\,\frac{v^2}{c^2} \qquad 2.66$$

By definition the kinetic energy is,

$$K = \tfrac{1}{2}\, m\, v^2 \qquad 2.67$$

If the particle was initially at rest, the change in kinetic energy is simply,

$$\Delta K = K = \tfrac{1}{2}\, m\, v^2 \qquad 2.68$$

Equating 2.66 and 2.68, we get

$$\tfrac{1}{2}E\,\frac{v^2}{c^2} = \tfrac{1}{2}\, m\, v^2 \qquad 2.69$$

Therefore,

$$E = mc^2 \qquad 2.70$$

Einstein had reasoned that if the kinetic energy of the particle is smaller by $\tfrac{1}{2}E\,\frac{v^2}{c^2}$, the only way this can happen is that the particle must lose mass when emitting radiation.

Here's an alternative, modern view of deriving the same equation using the 4-vector formalism developed at the end of section 2.1.

We set c = 1

We will measure the velocity of a free particle with respect to the proper time τ, not the ordinary time t.

Recall equation 2.33, reproduced below,

$$d\tau = dt/\gamma \qquad \qquad 2.71$$

And here γ from equation 2.14 is the same factor with c = 1,

$$\gamma = \frac{1}{\sqrt{1-v^2}} \qquad \qquad 2.72$$

The velocity of a free particle with respect to the proper time is defined as,

$$u^\beta = \frac{dx^\beta}{d\tau} \qquad \qquad 2.73$$

Using the chain rule and equation 2.71,

$$u^\beta = \frac{dt}{d\tau}\frac{dx^\beta}{dt} = \gamma \frac{dx^\beta}{dt} \qquad \qquad 2.74$$

Expanding the velocity vector into its 4 components,

$$u^\beta = (u^0, u^k) = \gamma \frac{dx^\beta}{dt}$$

$$= \gamma \left(\frac{dt}{dt}, \frac{dx^k}{dt} \right) \ where \ k = 1,2,3$$

$$= \gamma(1, v) = (\gamma, \gamma v) \qquad 2.75$$

The dot product between two vectors is now defined with the metric tensor (see equation 2.18),

$$u^2 = u \bullet u = \eta_{\alpha\beta} u^\alpha u^\beta \qquad 2.76$$

Note again that in u^2 the 2 means squaring, not the 2nd component of u.

Expanding the above into the temporal and spatial components,

$$u^2 = \eta_{00} u^0 u^0 + \eta_{ij} u^i u^j \qquad 2.77$$

Using equation 2.75,

$$u^2 = (-1)\gamma^2 + \gamma^2 v^2$$
$$= -1\gamma^2(1 - v^2) \qquad 2.78$$

But squaring equation 2.72 yields

$$\gamma^2 = \frac{1}{1 - v^2} \qquad 2.79$$

Therefore,

$$u^2 = u \bullet u = -1 \qquad 2.80$$

The 4-vector momentum is defined as mass times velocity, that is,

$$p^\alpha = mu^\alpha \qquad 2.81$$

Similarly, the 4-vector momentum has components,

$$p^\alpha = (p^0, p^k) \qquad 2.82$$

An important result is to calculate p^2, where again the 2 means squaring, not the component 2. First using equation 2.81,

$$p^2 = mu^\alpha mu^\alpha$$

$$= m^2 u^2 = -m^2 \qquad\qquad 2.83$$

Again expanding the above into the temporal and spatial components

$$p^2 = \eta_{00} p^0 p^0 + \eta_{ij} p^i p^j$$

$$= -1(p^0)^2 + 1(p^1)^2 + 1(p^2)^2 + 1(p^3)^2$$

$$-m^2 = -1(p^0)^2 + (\boldsymbol{p})^2 \qquad\qquad 2.84$$

Where $\boldsymbol{p}$ is the 3-vector momentum $= (p^1, p^2, p^3)$, and we used the result in equation 2.83 on the LHS.

We define $p^0 \equiv E/c = E$, (using the convention c=1) as the energy of the free particle.

Putting this altogether, we get,

$$E^2 = (\boldsymbol{p})^2 + m^2 \qquad\qquad 2.85$$

Putting the factor c back into the equation,

$$E^2 = (\boldsymbol{p}c)^2 + m^2 c^4 \qquad\qquad 2.86$$

For a particle at rest, the momentum is $\boldsymbol{p} = 0$. We get E = mc^2.

Conclusion: we can see the merit of the 4-vector formalism. But note that both derivations of Einstein's famous equation E = mc^2 were derived without a Minkowski diagram. And most importantly, the derivations are the result of the Lorentz transformation law, which itself is a result of the fact that the speed of light is an invariant in every inertial frame. It is the notion of the invariance under a Lorentz transformation that is carried into QFT.

2.8 Massless, Massive Particles and Tachyons

Consider now the Lagrangian for a free particle at rest in its own frame. As said above, p = 0 and E = mc². This means that the Lagrangian is $L = T - V = mc^2$

The action now becomes, (equation 2.39), reproduced below:

$$S = \int L\, dt \qquad\qquad 2.87$$

$$= mc^2 \int d\tau \qquad\qquad 2.88$$

Using dτ = - ds, we have

$$S = -mc^2 \int ds \qquad\qquad 2.89$$

The principle of least action demands that,

$$\delta S = 0$$

This means,

$$m\delta ds = 0 \qquad\qquad 2.90$$

$$mds = constant$$

Rewriting,

$$m = constant/ds$$

Consider just one spatial axis for simplicity, we have

$$m \; = \; constant/((ct)^2 \; - \; x^2)^{1/2}$$

For ct = x, the denominator is zero and so the only argument to avoid this infinity is that we have a massless particle traveling at the speed of light. For ct > x, we have a massive particle traveling less than the speed of light. Consider x > ct, the square root is now negative, and we have an imaginary number. We can do the following mapping,

$$mds \; \rightarrow \; imds$$

Where i is the imaginary number and ds is now a real number. Then equation 2.89 becomes,

$$S - - imc^2 \int ds \qquad\qquad 2.91$$

And this represents a particle with an imaginary mass. Such particles traveling faster than the speed of light are called tachyons in the literature. They have never been observed and are considered as unphysical.

We see again how the interval ds is linked to the Lagrangian, and the results one can deduce from such a relationship.

Here's a short and perhaps oversimplified version on the evolution of how physics has been done from the past to the present.

Galileo/Newton/Faraday: acceleration → forces → fields.

Euler/Lagrange/Hamilton: forces → energy → Lagrangian & Hamiltonian → equations of motion (eom).

Planck/Einstein/Bohr/Schrödinger/Heisenberg/Dirac:
quantized energy → Schrödinger's equation (eom).

Modern view: Lorentzian & gauge invariant Lagrangian →
fields & particles.

One cannot overemphasize the importance of the
Lagrangian.

"Every law of physics, pushed to the extreme, will be found to be statistical and approximate, not mathematically perfect and precise."

- John Wheeler (1911-2008)

Chapter 3

New Insights into the Concept of Entropy

(A New Law of Kinematics Is Introduced)[3][4]

Upon examination of the concept of energy, two universes are possible: 1) energy can be infinitely divisible and the universe will suffer the well-known heat death; 2) energy is not infinitely divisible, that it is quantized, and the universe does not suffer a heat death.

One version of Zeno's famous paradox is that you can't cross the street from one side to the other because you need to cross one half of the distance, then one half of that half, then one half of the one half of that half, and so on. In the physical world this is resolved very quickly. As you are crossing each one half, you will come to the situation that your feet will be longer than whatever remaining distance you need to cross. Similarly with energy, you can exchange any amount of energy until you reach the smallest quantity of energy, a particle, which cannot be further divided. So how are we going to distinguish case 1 and 2, and justify why energy is quantized? A new law of kinematics can resolve this issue.

In a toy model, a new law of kinematics, named for reasons explained later as the 3rd Law of kinematics leads to the concept of energy being quantized. It also offers an explanation of why entropy increases in that universe.

Toy models play a vital role in physics. It is well-known that the de Sitter model in cosmology plays an important role in physical applications [5]. Three laws of kinematics are identified in this toy model. The first two were already known. The third law of kinematics is a new law of kinematics in a universe of bouncing balls in which there is no exchange of angular momentum (no spin involved), and it states: given an inertial frame, that for free particles, under no circumstances a body with higher kinetic energy can gain energy from another body with lower kinetic energy. Stated differently: in an elastic collision for a given inertial frame, it will always be the case that the higher kinetic energy body will always lose energy to the lower kinetic energy body, and the lower kinetic energy body will always gain energy from the higher kinetic energy body. Moreover, under general considerations from the theory of Special Relativity, a particle can decay into other particles only if there is a source of positive energy necessary to insure that the new particles have positive kinetic energy, which is why $E=mc^2$ provides the source of this positive energy in the form of mass conversion into energy. Also this new kinematics leads to the reformulation of the 2nd law of thermodynamics which explain why heat can only flow from a hot body to a cold body, and never the other way around, that a system left on its own will see its disorder increase in time, and that the entropy as defined according to Boltzmann tends to increase. The insight that we gain from this deeper concept which underlies the new law of kinematics and the reformulation of the second law of thermodynamics is that vacuum energy is a necessary condition for particles to pop out of the vacuum. However, this spells trouble for Hawking radiation as in the way it was formulated, it would lead to a violation of this new kinematics law. Surprisingly this new law of kinematics also sheds new insights into Planck's ad hoc hypothesis, $E = \hbar\omega$, and why energy must be quantized.

Note: This new law of kinematics could have been discovered at any time after 1905 when Einstein published his paper on Special Relativity. Why it wasn't discovered until recently still baffles me to this day. We will call it the 3rd law of kinematics. The first two are already known, and have been known under different names, which we will now review.

3.1 First and Second Law of Kinematics

First law of Kinematics (a.k.a. Galileo's law of inertia or Newton's 1st law of motion)

Galileo's inertial law of motion, ignoring the curvature of the earth, can be represented as one body viewed in two different frames of reference.

Frame 1   Frame 2

Fig. 3.1

We have two observers: one is at rest with the body, Frame 1; the other observer is in uniform motion with respect to the same body in question, Frame 2. In both of these frames, the object experiences no net force.

Second Law of Kinematics

This law was also known initially as Galilean Transformation Law, which was subsequently replaced by the Lorentz transformation laws we saw in chapter 2. It is concerned with how in different inertial frames of reference such as Frame 1 and 2 the laws of nature are transformed. Initially, we had the Galilean transformation equations:

$$x_2 \rightarrow x_1 - vt \qquad\qquad 3.1$$

In the theory of Special Relativity (SR) Einstein made us all aware that in order to have the constancy of the speed of light in every inertial frame of reference, the Galilean transformation equations had to be replaced by the Lorentz transformation equations:

$$x_2 \rightarrow \gamma(x_1 - vt) \qquad\qquad 3.2$$

$$t_2 \rightarrow \gamma\left(t_1 - vtx_1/c^2\right) \qquad\qquad 3.3$$

Where again we have $\gamma = \left(1 - \dfrac{v^2}{c^2}\right)^{-\frac{1}{2}}$

As it was noted before, this guarantees that the speed of light is an invariant.

The well-known result of equations 3.2 and 3.3 is that time is no longer an invariant quantity but when measured in a given frame it will yield a different time measurement compared to another frame, and this effect is generally known as time dilation (section 2.4).

The argument that the proponents of a 4-D universe is that time t gets mixed up with position x in the Lorentz transformation equations. And therefore it acts as another coordinate. However, as we have argued in section 2.3 space-time intervals on a Minkowski diagram are to be dealt with extreme caution as it is more of a graph than a coordinate system. Therefore one should think of equation 3.3 as,

$$ct_2 \rightarrow \gamma \left(ct_1 - \frac{v}{c}x_1\right) \qquad\qquad 3.4$$

It's just that one shouldn't forget that not only do we get time dilation, but also length contraction, so our length ct_2 travelled by light as measured in the second frame will be shortened.

Now we will formulate a new law of kinematics, the third law of kinematic, which brings new insights particularly into thermodynamics.

3.2 The First Law of Thermodynamics

The 1st law of thermodynamics states that the total energy of a closed system is always conserved. But it does not specify how the energy is redistributed after an elastic collision. Denote the higher kinetic energy particle as B-particle, B ≡ "Big", and the lower kinetic energy particle as

L-particle, L≡ "Little". The 1st law of thermodynamics states after an elastic collision,

$$KE_B + KE_L = KE'_B + KE'_L \qquad 3.5$$

Where $KE \equiv$ Kinetic Energy before collision, $KE' \equiv$ Kinetic Energy after collision.

3.3 A New Law of Kinematics

The most simple collision one can think of – no spin, no deformations, no fields involved – is an elastic collision.

Now this new law of kinematics states that,

$$KE_B > KE'_B \qquad 3.6$$

That is, the B-particle always loses kinetic energy after the elastic collision. Rewriting equation 3.5 as,

$$KE_B = KE'_B + KE'_L - KE_L \qquad 3.7$$

Substitute (3.7) into (3.6),

$$KE'_B + KE'_L - KE_L \; > \; KE'_B$$

$$KE'_L - KE_L \; > \; 0$$

$$KE'_L \; > \; KE_L \qquad 3.8$$

Consequently, after an elastic collision the corresponding L-particle always gains kinetic energy after an elastic collision.

We can justify this law (equation 3.6) with the help of Einstein's law of Relativity by considering different observers in different inertial frames of reference.

3.4 Special Relativity

As it was mentioned in the previous chapter, at the core of the theory of Special Relativity is that the laws of physics are the same in all inertial frames. We will take a second look at this with the following illustration.

In case A, we have an observer which is at rest with the L-particle. In case B, we have an observer which is not at rest with the L-particle. We refer to the latter case as the "Lab" frame.

Before Collision

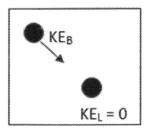

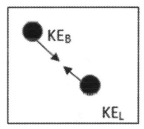

Case A: $KE_L = 0$ Case B: $KE_B > KE_L$

Rest frame   Lab frame

Fig. 3.2

After Collision

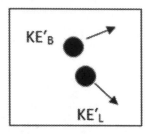

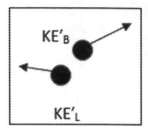

Case A: $KE'_L > 0$

Case B: $KE_B > KE'_B$

Rest frame

Lab frame

Fig. 3.3

We make the following claim that a particle can never have negative kinetic energy, which is what is observed in Case A after an elastic collision. Therefore in the rest frame, the L-particle must have $KE'_L > 0$, (Case A).

So if we compare fig 3.2 and 3.3 in the frame of an observer at rest with respect to the L-particle (case A), we see that the L-particle has gained kinetic energy after the collision. By conservation of energy, the B-particle must lose energy after an elastic collision ($KE_B > KE'_B$), which is equation 3.6.

This is the new law of kinematics, the 3rd law of kinematics: given an inertial frame, in every elastic collision, it will always be the case that after an elastic collision the higher kinetic energy body will lose kinetic energy and the lower kinetic energy body will gain kinetic energy. Stated differently: in every frame of reference, the

B-particle loses kinetic energy, and the L-particle gains kinetic energy. There are no exceptions.

Section 3.5.1 will review these collisions, taking into consideration the invariance of the change in kinetic energy. But next we will examine in what ways this new law leads to a reformulation of the 2nd law of Thermodynamics.

3.5 Heat Transfer from Hot to Cold

There are alternative views of looking at the concept of entropy. We shall use water in its different states (ice to water to vapor) to illustrate these points of views.

One of these is the concept of order/disorder. A simplified version of this argument goes as such: in ice we have a lot of order, that is, each molecule of ice is frozen in place, forming a crystal structure. As the temperature rises, and the ice is turned into liquid then vapor, the molecules are moving more and more at random. Hence we have less order, and greater disorder. The increase in entropy is seen as an increase in disorder.

A second view of entropy is seen through the concept of information availability. The argument goes as such: in ice, the molecules being frozen are really vibrating slowly about some position in a crystal form. The availability of that information means that it's easy to keep track of the position of each molecule. As the temperature rises, and the molecules get more and more agitated, it is harder and harder to keep track of those molecules. The increase in entropy is seen as information being less available – not to confuse with information being lost, which is another issue.

Now we turn our attention to what entropy was originally thought of: heat flows from a hot body to a cold body. The ice melts down because heat from the environment, which is at a higher temperature, flows into the ice, the colder body. It is this point of view we will consider in our reformulation of the 2nd law of thermodynamics.

A hot body is a system in which the average kinetic energy of its constituents is high (high temperature), whereas a body is considered cold with respect to the hot body if it has a lower average kinetic energy (low temperature). When these two systems are placed in contact with each other, the new law of kinematics says that in every collision, B-particles always lose energy, while L-particles always gain energy. Given sufficient time, the average kinetic energy of all the B-particles will decrease, while the average kinetic of all the L-particles will increase. This is a manifestation of heat flowing from a hot body to a cold body.

Let us demonstrate why this is so.

Consider that for a hot body, the system is made up of B-particles. For every collision taking place when the two systems are in contact, we get the following, (each particle is labelled B_1, B_2, B_3,... B_n)

$$KE_{B1} > KE'_{B1} \qquad\qquad 3.9$$

$$KE_{B2} > KE'_{B2}$$

.

$$KE_{Bi} > KE'_{Bi}$$

We get,

$$\sum_1^n KE_{B_i} > \sum_1^n KE'_{B_i} \qquad\qquad 3.10$$

The last step was obtained by summing up over all the B-particles. But we're missing one element to work out the average kinetic energy. The temperature is a measure of the average kinetic energy, and so we need to divide the sum by the number of particles. Once the two systems are in contact, the question is, will the number of B-particles remains constant, or will it change? To answer this, we will use the following illustration. We will consider a system of three particles where $KE_1 > KE_2 > KE_3$, (Fig. 3.4))

Case	Collision	KE_1	KE_2	KE_3	N_0	N_L	Total Energy	Ave. Total Energy
1	none	30	20	10	1	1	60	20
2	1↔2	29	21	10	1	1	60	20
3	2↔3	29	20	11	1	1	60	20
4	1↔3	27	20	13	1	1	60	20
5	1↔2	26	21	13	1	1	60	20
6	2↔3	26	19	15	1	1	60	20
7	1↔3	24	19	17	1	1	60	20
8	1↔2	23	20	17	1	1	60	20
9	2↔3	23	19	18	1	1	60	20
10	1↔3	21	19	20	1	1	60	20

Fig. 3.4

(a) In case 1, we have no collision. This is our starting point. The energy assigned to the particles are arbitrary and will be sufficient to illustrate what is happening just by applying the new law of kinematics. So we have one B-particle, particle 1 (N_B= 1), and one L-particle, particle 3 (N_L= 1). We ignore particle 2 for reasons that will be obvious.

(b) We notice that particle 2 (cases 1 to 9) with a kinetic energy close to the average kinetic energy (Total energy = 60, Ave. KE = 20) fluctuates around the average. When it interacts with particle 1, it is the L- particle, and so it gains energy (cases 2, 5, 8). When it interacts with particle 3, it is the B-particle, so it loses energy (cases 3, 6, 9). So we don't count it in the N_B column nor in the N_L column.

(c) Particle 1 is always the B-particle, so it loses energy in every collision.

(d) Particle 3 is always the L-particle, so it gains energy (except for case 10, which we will discuss later). The net result is that for all particles, their kinetic energy tends towards the average kinetic energy.

(e) In case 10, the last entry, particle 2 has less energy than particle 3. We just need to relabel, 2 → 3, and 3 → 2. And so N_B= 1 and N_L= 1.

In light of the reformulation of the 2nd law, we can look at our B-particles, in the system we've designated as the hot body, as particles wearing a tag which reads "B". Similarly, for the cold system, the particles are wearing tags with the label, "L". So now we let them loose in the same room. They start bumping into each other. Suppose in that process that one of the L-particles has gained enough energy so that it has earned to be part of the B-team. Call it the "lucky" particle. All we need to do is switch tags: there is at least one B-particle that has less energy than lucky particle, otherwise our lucky particle isn't "lucky". After the switch, the number of B-particles and L-particles, both remain the same. We can do that for every single collision: either we switch tags or we don't.

We can safely say that the number of B-particles (N_B) remains constant, as well as for the L-particles (N_L). So to calculate the average kinetic energy, we add the energy of each particle divided by the number of particles. Equation 3.10 becomes,

$$(\Sigma\ KE_{Bi})/\ N_B > \Sigma\ KE'_{Bi}\ /\ N_B \qquad\qquad 3.11$$

Or

$$Ave.\ KE_B > Ave.\ KE'_B \qquad\qquad 3.12$$

And this can only hold if N_B is the same throughout the process when the two systems are placed in contact, which our table demonstrates (Fig. 3.4).

3.5.1 Center of Kinetic Energy

We know that velocity is not an invariant quantity as we transform our equation for different inertial frames, and therefore neither is the kinetic energy. However, as Einstein assumed in his derivation of $E = mc^2$ (see section 2.7), the change in kinetic energy is invariant. We will illustrate that situation but first some definitions.

Definition: The center of kinetic energy is that frame of reference in which both particles, the B-particle and the L-particle, would have their kinetic energy equal to the average kinetic energy.

Note: The center of kinetic energy is an idealized frame of reference in which $KE_B = KE_L$.

<u>Definition</u>: The deviation is the value of the energy difference between the body's kinetic energy in a given frame and its kinetic energy in the frame of the center of kinetic energy.

Irrespective of how we define our energy scales according to one frame of reference, what matters is that we measure energy differences. And this difference, or the deviation, is frame invariant.

So the question arises: consider the B-particle, which we claim that it will always lose energy in a given collision, could it loses all of its energy to the L-particle?

<u>BEFORE COLLISION</u>

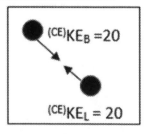

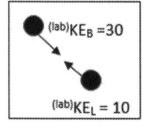

(a) $KE_B = KE_L$ = Ave. KE
Center of Energy

(b) $KE_B > KE_L$
Lab Frame

Fig. 3.5

Going back to fig. 3.4, where case 1 is the starting point with $^{(lab)}KE_B$ =30 and $^{(lab)}KE_L$ = 10, this is illustrated in Fig. 3.5b; while Fig. 3.5a depicts the same situation for an observer in the frame of reference which is the center of kinetic energy. For this case, we have $^{(CE)}KE_B$ = 20 and $^{(CE)}KE_L$ = 20. In terms of the deviation we write for the B-particle,

$$\| \, ^{(lab)}KE_B - \, ^{(CE)}KE_B \, \| = \text{Dev. KE} \qquad\qquad 3.13$$

For the L-particle,

$$\| \, ^{(lab)}KE_L - \, ^{(CE)}KE_L \, \| = - \, \text{Dev. KE} \qquad\qquad 3.14$$

Using the values for fig 3.4, we obtain

$$\text{Dev. KE} = 10 \text{ for B}, -10 \text{ for L} \qquad\qquad 3.15$$

AFTER COLLISION

Suppose that the B-particle loses all of its energy to the L-particle, Fig. 3.6b. Then what we would see after collision, (after applying equations 3.13, 3.14 and 3.15) is depicted in Fig. 3.6a.

Note that the average kinetic energy hasn't changed, but KE'_B has now a negative value in that frame, and this is forbidden. And so this is an unphysical situation.

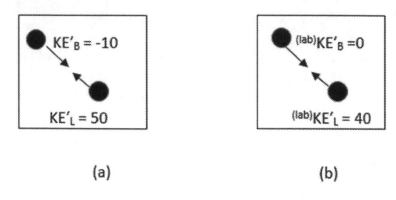

(a)   (b)

Fig. 3.6

After COLLISION (considering thermal equilibrium as the limiting process)

In this case, the B-particle loses its maximum energy, which is restricted by the thermal equilibrium set at KE_{ave} = 20 (Fig. 3.7).

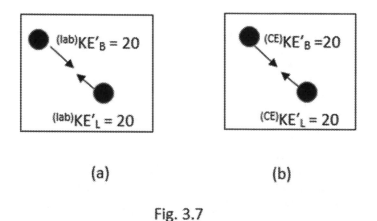

(a)   (b)

Fig. 3.7

Note that the lab frame is now the equivalent of the center of energy as the B-party (L-particle) has lost (gained) the maximum to bring it to the average kinetic of the system, and no particle has negative kinetic energy.

And so, we can conclude that the B-particle will lose energy in the lab frame, but will never lose all of its energy. The maximum it can lose is up to the value that reaches the average kinetic energy at the thermal equilibrium.

3.6 Thermal Equilibrium

In standard interpretation of thermodynamics, equilibrium is defined in terms of an isolated system which is found with equal probability in each one of its accessible states[6]. We now redefine thermal equilibrium as a limiting process. We need to re-examine the table in Fig. 3.4 and compare each particle's kinetic energy with the average kinetic energy of the combined system. We notice that the difference of the B-particle's kinetic energy with the total system's average energy tends to decrease towards the average kinetic energy of the combined system. We can express that as,

$$\text{Ave. } KE_B \rightarrow \text{Ave. } KE \text{ of the combined system} \qquad 3.16$$

And also, the difference of the L-particle's kinetic energy with the total system's average energy tends to increase to the same average.

$$\text{Ave. } KE_L \rightarrow \text{Ave. } KE \text{ of the combined system} \qquad 3.17$$

Since temperature is a measure of the average kinetic energy for any given system, for the combined system in our case, this temperature is the equilibrium towards which the kinetic energy of both the B-particles and the L-

93

particles are moving. The result is that the number of B-particles (and the L-particles) remains constant. When the combined system has finally reached thermal equilibrium, the B-particle (the L-particles) being near the equilibrium will lose (gain) kinetic energy sufficiently, requiring a frequent change of tags, nevertheless, the collisions even as tiny these differences in kinetic energy can get will continue relentlessly, which can be observed as fluctuations. Thus, a thermometer sensitive enough will be able to show those fluctuations.

The combination of the 1st and 2nd laws of thermodynamics makes this process necessary, and not just probable. This is a manifestation that heat flows from a hot body to a cold body such that the combined system will eventually reach thermal equilibrium.

3.7 Positive Definite

So the question arises, how does this reformulation fit in the scheme of Boltzmann's microscopic states? Boltzmann had to assume that all those states were equally probable, and so it necessitated the use of probability theory in which each outcome is a number between 0 and 1, and the sum of all outcomes is 1. What this means is that Probability theory uses the concept of positive definite – you can't have negative probabilities. It turns out that kinetic energy is also positive definite, (see equation 2.67). If we look again at Fig. 3.4, Boltzmann assumed that the B-particles could equally take any value between 0 and 60; similarly with the L-particle, any value with equal probability between 0 and 60 as long as the law of conservation of energy is obeyed. What the new law of kinematics says is that the B-particle in that particular frame of reference can take values with equal probability but only between 20 and 30, while the L-particles can only take those values with equal probability only between 10 and 20.

Needless to say that in a different frame of reference these ranges of values would change. In case A above (section 3.4), the L-particle being at rest would take values of energy ranging between 0 and 20 after collision, while the B-particle would be in the range 20 to 40 in the aftermath of the collision. In case B, those numbers would be reversed. So we can see that the number of microscopic states available to each particles is restricted and depends on the frame of reference.

Though Boltzmann's work was a step in the right direction, his analysis of ensemble of particles obscured what was happening at the individual level, where one particle collides with another particle. And only by taking into account this new law of kinematics and a reformulation of the 2nd law can bring to the fore why heat always flow from hot to cold.

We can now claim that we've inherited all the mechanism of what has been established in thermodynamics.

Now let's revisit the familiar case of an ice cube left in the open, and finding out later on it has evaporated. The standard description is that the atoms in the cube gain energy from the air molecules sufficient enough to escape their bondage in the cubic crystal. This is the interpretation that as the entropy increases – that is, the atoms arranged as cubes in a certain order but later on, as free particles in the air – they have less order, or greater disorder. This interpretation has led to the following misconceptions that can now be dispelled with the new law of kinematics:

(i) There is the belief that some molecules in the air could join the water molecules in the cube, but that the probability for that to happen is so small, so insignificant that we can ignore that.

(ii) We are also familiar with the scenario often repeated that the air in the room could all gather in one corner, leaving everyone in the room gasping for air. But have no such fear we are told, not to worry that for this to happen, it is very unlikely.

With the reformulation of the 2nd law we can say that the air molecules are the B-particles in the room, those in the ice cube are the L-particles. The average kinetic energy of the B-particles tends to decrease to some equilibrium. Unless the outside temperature is well below zero centigrade, the temperature at equilibrium for this case, the air molecules in their tendency to lose kinetic energy up to thermal equilibrium will still have enough energy to not solidify. So the air molecules freezing to join the ice cube is not going to happen.

In the case of the air molecules all moving towards the corner of the room, we would observe an increase in temperature in that corner, and for that to happen, the B-particles which have kinetic energies slightly above thermal equilibrium would have to gain energy, and we know this is forbidden by the new law of kinematics. Unless there is a force shooing in these particles into one single corner, no such phenomenon is going to happen.

3.8 E = mc² Revisited

Consider a particle at rest that decays into two particles,

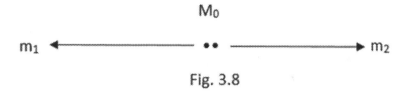

Fig. 3.8

Initially, the total energy is zero. After it decays into two particles, they will fly away from each other with equal and opposite momentum (conservation of momentum) and with equal and opposite spin (conservation of angular momentum). But they both have kinetic energy, and by the conservation of energy,

$$0 = KE_1 + KE_2 \qquad\qquad 3.18$$

This necessitates that one of the particles would carry NEGATIVE kinetic energy?! However according to the new law of kinematics, this is forbidden. The only way this can happen is that some positive energy is given to these two particles, and Einstein provided the answer in E= mc², that is, there is a mechanism by which this process can take place. I will make this statement stronger: It would be impossible for decay to occur in nature if mass could not be converted to energy. What Einstein discovered, although admittedly by other means, is that the existing laws of kinematics had to be amended in view of the constancy not only of the speed of light but all laws of physics in all inertial frames by making everyone aware that the Galilean transformations must be replaced by the Lorentz transformations – an idea that was successfully incorporated in Quantum Mechanics to yield Quantum Field Theory. Einstein revised the above equation to read as,

$$(\text{Mass, } M_0 \rightarrow \text{energy}) = KE_1 + KE_2 \qquad 3.19$$

More specifically, the above equation is worked out to give for one particle (equation 2.85 with c=1),

$$E^2 = p^2 + m^2 \qquad 3.20$$

In the case of a particle of mass M_0 decaying into two particles of mass m_1 and m_2, we get

$$M_0^2 = p_1^2 + m_1^2 + p_2^2 + m_2^2 \qquad 3.21$$

Rearranging,

$$M_0^2 - (m_1^2 + m_2^2) = p_1^2 + p_2^2 \qquad 3.22$$

The condition for decay to take place and making sure it obeys the new kinematics, that is, no particle can have negative kinetic energy in any frame of reference is,

$$M_0^2 > (m_1^2 + m_2^2) \qquad 3.23$$

The mass can change as long as it obeys this restriction. The implication is a mass can decay through different channels, an observation that has been confirmed multiple times in high energy physics. However a re-interpretation is necessary in light of the new kinematics: whenever new particles appear, there must be a source of energy such that no new particle can have negative kinetic energy. A corollary to this is that nuclear decay is permissible in nature because matter can be converted to energy. Einstein's discovery, $E = mc^2$, is an expression of this corollary.

3.9 Vacuum Energy

It was Schrödinger who came up with the Klein-Gordon equation (see sec 4.11, below) in his search to describe de Broglie waves. But it gave out negative frequencies and negative probabilities. Hence he abandoned it and used a non-relativistic approximation in his work. Dirac was able to give an explanation in terms of a negative sea with negative particles and the condition that it was all filled up. Occasionally a hole would respond to electric fields as though it were a positively charged particle and predicted the existence of anti-matter, which was discovered subsequently a few years later. Today we treat the positron as a "real" particle rather than a hole or the absence of a particle, and the vacuum is thought as the state in which no particles exist instead of an infinite sea of particles. On the other hand quantum field theory postulates that at every point in space, there is a field made of an infinite number of harmonic oscillators, which are expressed as Fourier series and annihilation and creation operators. With what has been established in this chapter, we can say that it is only through this device of a zero-point energy that quantum fluctuations are a possibility. So even though some are uncomfortable with this infinite zero-point energy, it is a necessity as it is a testimony that the new law of kinematics must prevail: no particles can pop out of the vacuum without a source to supply it with energy sufficient enough so that these particles can escape with positive kinetic energy. What remains to be determined is how big this ground zero energy is as it is at variance with Dark Energy by the most mismatched ratio in science history. Hopefully the new law of kinematics and this reformulation of the 2nd law of thermodynamics are steps in that direction.

3.10 Feynman Propagators

In retrospect, Feynman's contribution to the concept of propagators in the context of Quantum Field Theory (QFT) fits in well within the 3rd law of kinematics and what was said in section 3.8.

We will not delve too deeply into QFT as it is beyond the scope of this book. However, we will draw briefly the main lines as to what propagators are and what they do.

Historically Quantum Mechanics (QM) was developed approximately between 1900 and 1927, while QFT's development took place between 1927 and 1971. Two basic reasons for that underlined development is: (1) Special Relativity had to be incorporated into QM; and (2) the realization that particles could be created and annihilated. In the first case, the position operator x was replaced by a field operator $\phi(x)$, in which the position x is a parameter, putting it on equal footing with time as a parameter (chapter 2). In the second case, the amplitude (see sections 4.11) is redefined from $< \psi \mid \psi_i >$ (equation 4.21) to a propagator of the type,

$$G\,(x, t_x, y, t_y) = \theta(t_x - t_y) < x(t_x)\mid y(t_y) > \quad 3.24$$

The theta function $\theta(t_x - t_y) = 1$ if $t_x > t_y$. Otherwise it is zero. The function G is a Green function[7].

Equation 3.24 is thought of as a particle being annihilated at position y, and then created at position x (t_x occurs later than t_y). Hence, a particle is "propagated" from y to x. (See Fig. 3.9)

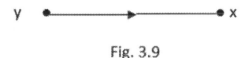

Fig. 3.9

Fcynman struggled over how to include antiparticles within the context of particle creation/annihilation into the propagator, and his original thought was to conceive of an antiparticle as moving backward in time[8]. Today we think of charged particle flow – the antiparticle having the opposite charge as describing a flow of charges in opposite direction to its particle. Fig. 3.10 describes this situation with a photon annihilated at y with the creation of an electron and its antiparticle, the positron. At position x, the reverse takes place: both the electron and positron are annihilated with the creation of a photon.

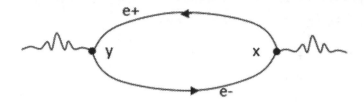

Fig. 3.10

This is quite similar to Fig. 3.8 which depicts a particle M_0 decaying (annihilated) into two particles m_1 and m_2 (created). The only way this can happen is that along with the energy, momentum is conserved, and that requires the creation of at least two particles – something that can't happened in Fig. 3.9. One can say that by including the antiparticle in the calculation of the amplitude, Feynman was anticipating the 3rd law of kinematics.

3.11 Hawking Radiation

We will discuss the case of Black Holes in a later section. For now, we will consider Black Holes as theoretical possibilities.

What doesn't fit in well within the 3rd law of kinematics and what was said in section 3.8 is Hawking radiation.

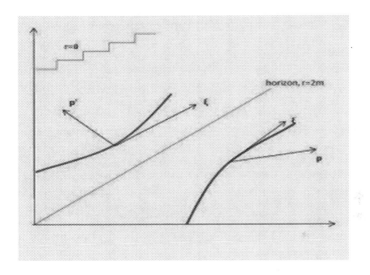

Fig. 3.11

Fig 3.11 shows a rest-mass zero particle-antiparticle pair which has been created by vacuum fluctuations in such a way that the two particles were created on opposite sides of the event horizon of a black hole. The vector ξ is known as a Killing vector. The argument states that the components $\xi \bullet p$ and $\xi \bullet p'$ must be equal and opposite so that $\xi \bullet (p + p') = 0$, (value of the vacuum). The particle ($\xi \bullet p > 0$) can propagate and can be seen as radiation by an observer at infinity. This also means that the antiparticle ($\xi \bullet p' < 0$) will be absorbed by the black hole, thus decreasing its mass in the process. This is the basis of Hawking's claim that black holes radiate, and in time, will evaporate.

Hawking [9][10] assumed that particles can pop out of the vacuum. As it was pointed out above, this process is permissible as long as we take the vacuum energy as necessary to allow this process. And this process has been used successfully in the Casimir force and the Lamb shift, to name two instances. In the case of Hawking radiation, it was argued that the Black Hole loses energy by absorbing one of the free particle with negative kinetic energy, and the other free particle just outside the Horizon escapes to infinity [11]. To an observer at infinity, she would see those particles coming at her, while the Black Hole loses mass. In the process, the Black Hole radiates. And Hawking formulated that its mass would be inversely proportional to its temperature. This is all well within the 1st law of thermodynamics.

There are some important issues with Hawking radiation that seem to be irreconcilable with known principles:

(i) Negative energy is counter to the very basic notion in QM that a ground state – a state of minimum positive energy – must exist otherwise a particle with negative energy would slide irreversibly to negative infinite energy. In other words, no ground state means no QM.

(ii) As it was mentioned in section 3.8, there must be a source of energy to give particles created in such a process some positive kinetic energy. If that energy comes from the vacuum, then the equation $\xi \bullet (p + p') = 0$ must be replaced by $\xi \bullet (p + p') = \Delta E_{vacuum}$ (c=1). In other words, the vacuum would be losing energy. In this process of Black Holes radiating energy, the vacuum would be depleted everywhere there are Black Holes doing exactly that. And the Black Hole would gain in mass, not lose its mass, and accordingly, its temperature would decrease.

(iii) An alternative is to have tachyons (imaginary masses, section 2.8). These would be travelling faster than light. But no such particle has ever been observed.

As it stands, Hawking radiation is problematic. Notwithstanding this puts a hole in the Information Paradox[12] and the Firewall Controversy[13].

3.12 Quantized Energy

The third law of kinematics makes it imperative that energy at sub-microscopic scales is quantized.

In the exchange of kinetic energy between the B-particles and the L-particles, as it was mentioned before, this leads to thermal equilibrium and produces fluctuations. But why should this process stop there? Why not go all the way until all the B-particles (L-particles) have lost (gained) all of their energy permissible, in which case their kinetic energy would then equal exactly to the average kinetic energy? If that would happen, then all of the B-particles and all the L-particles are identical, in that, they all have equal kinetic energy, the average kinetic energy of the combined system, and no exchange of energy is ever possible. In this universe, there are only bouncing balls, in which all the B-particles and all the L-particles are the same. Call that the heat death. But that's not the real world. The only way out is that there exists a minimum non-zero energy in which the B-particles can still exchange with the L-particles, and in every single collision we either exchange tags or we don't. So we can say that,

$$E_{min} = k(?)$$ 3.25

Where the k is a constant that represents some minimum value, and the question mark (?) represents a quantity to which the energy can be related.

And so what is/are the candidate(s) possible suitable for that unknown quantity?

3.12.1 The Harmonic Oscillator

Quantum Field Theory (QFT) began with the works of Dirac and others, and as stated before, the key idea was to unify Quantum Mechanics (QM) with SR. By demanding that the Lagrangian is Lorentz invariant (the 2nd law of kinematics), it guarantees that SR is incorporated into the theory. From QM, what was used essentially is the quantized harmonic oscillator – the only problem with an exact solution to the Schrödinger's equation – which contains everything you need in QM.

In our previous assumptions, it was stated that no energy enters our system, no energy leaves our system. We just have free particles. It was also stated that if a particle breaks up into two or more particles, a source of energy had to be available to give each particle positive kinetic energy, and that was the mass in the form of $E = mc^2$. In an elastic collision what we're dealing with is not a break-up, but just an exchange of energy. So mass as a possible candidate for our unknown quantity is ruled out. Consider now that our particles are really harmonic oscillators, the only remaining candidate would be their frequencies, and that's what oscillators do – they oscillate at a certain frequency. So we can write equation 3.25 as,

$$E_{min} = k\omega \qquad\qquad 3.26$$

This would be the minimum quantity of energy that must be exchanged in order to avoid the heat death. But what still remains to be shown is that the energy is quantized.

Let us assume that in any elastic collision, the energy exchanged can be expressed as,

$$E_{exchanged} = nE_{min} = nk\omega \qquad\qquad 3.27$$

All we need to do is to show that n is an integer.

3.12.2 Normal Modes

And now we go back to Planck and the experiment he needed to explain black body radiation. What was used in those days is that radiation inside a black box was made up of frequencies of all modes. That was permissible as Fourier had successfully demonstrated that such frequencies can be expressed as a series of fundamental modes, also called normal modes. More specifically, we will be concerned with the wavelength,

$$c = \lambda v \qquad\qquad 3.28$$

Multiply both sides by 2π, and $\omega = 2\pi v$, rearranging,

$$\omega = 2\pi c/\lambda \qquad\qquad 3.29$$

But these normal modes obey boundary conditions in the box, such that the 1st mode is $L = \frac{1}{2}\lambda$, the 2nd mode is, $L = \lambda$, the third, is $L = 1\frac{1}{2}\lambda$... $L = (n +\frac{1}{2})\lambda$, where n is an integer. Substitute that into equation 3.29,

$$\omega_{quantized} = (2\pi c/L)\ (n + \tfrac{1}{2}) \qquad\qquad 3.30$$

With this result, we can repackage equation 3.27 as,

$$E_{exchanged} = nk\omega = (n + \tfrac{1}{2})\hbar\omega \qquad\qquad 3.31$$

Where $\hbar$ is a universal constant known as the Planck constant (in some text, the reduced Planck constant), and the energy is quantized.

3.12.3 Hidden Assumptions Uncovered by QM

We see that Quantum Mechanics uncovered a hidden assumption. When Newton put physics on the map with his three laws of physics and his enterprise was further refined later on with concepts such as momentum, energy, and entropy, which in effect started a project now call Classical physics, it was believed throughout those years that energy could be infinitely divisible. Planck discovered this wasn't so in the spectrum of black body radiation, in effect that energy at atomic and sub-atomic scales behaved differently. Einstein put the right interpretation by saying that light came as "quantized package' that behaved like particles, which we now call photons, in the emission and absorption of light by matter.

We now see that the heat death so feared in the 19th century due to thermodynamics considerations can be averted with the 3rd law of kinematics demanding that energy being quantized.

3.13 Irreversibility and the Arrow of Time

As we have mentioned in chapter 1, the arrow of time is a mental construct. What we have is "matter moving through space". Or we simply move. The typical argument is the movie of a window pane made of glass, falling from a certain height, shattered into a multitude of broken pieces, and then showing the movie in reverse depicting all the pieces coming together to form back the window pane and climbing back to the same height. We're supposed to ask: how come we never see that in nature??? What we forget in this scenario is that the window pane was shattered because a force was acted upon it. To get the window pane back from the shattered piece would necessarily require a force to do that exactly, which is quite complicated as initially sound was produced and propagated through the air, heat was released to the environment, and the chemical bondings between molecules were broken. Reconstituting the window pane is not an easy task, but it can be done since the window pane before it was broken was manufactured in the first place.

With the third law of kinematics we can add one more argument. Consider the case discussed in section 3.5, in which we consider two bodies – one hot body made of B-particles, and the other cold, made of L-particles. Suppose all the particles in these two bodies are identical in mass, in other words, they can only be distinguished by their differences in kinetic energies. In every collision, the B-particles lose energy and the L-particles gain energy (3rd law of kinematics). Suppose someone played the movie in reverse without telling us. We would observe that the B-particles are gaining energy, and the L-particles losing energy in every collision. This would mean that in the rest frame of the L-particles, these would need to have negative kinetic energy after collision, which is impossible. We would now know that this movie is being played in reverse. This illustrates the breaking of time symmetry, another consequence of the 3rd law of kinematics.

3.13.1 CPT Theorem

In view of what was said in chapter 1 about time and the new law of kinematics implying that time is no longer symmetrical, how does that impact the CPT theorem in QFT?

In QFT, an important result is that the Lagrangian is invariant under a CPT transformation. Let's define what these objects are. The "C" is charge conjugation. Under such transformation, we turn a particle into its antiparticle (same mass, opposite charge):

$$C: \quad e^+ \rightarrow e^-$$

The "P" stands for parity. It transforms the position x into − x.

$$P: \quad x \rightarrow -x$$

While "T", the time reversal operator, transforms time t into −t.

$$T: \quad t \;\rightarrow\; -t$$

The condition that in QFT the Lagrangian is invariant under Lorentz transformation and unitarity demands that the complex number i is also turned to −i [15]:

$$T: \quad i \;\rightarrow\; -i$$

Because the momentum is (equation 4.22, in chapter 4): $\rightarrow -i\,\hbar\frac{\partial}{\partial x}$, the operator T flips the momentum. It does likewise for the spin.

Another look at this is through the Schrödinger equation (equation 4.24 in chapter 4):

$$i\hbar\,\frac{d\,|\psi>}{dt} \;=\; H\,|\psi>$$

Where H is the Hamiltonian operator. Applying the time reversal operator to that equation, we get:

$$T: i\hbar\,\frac{d\,|\psi>}{dt} \;=\; H\,|\psi> \;\rightarrow\; -i\hbar\,\frac{d\,|\psi>}{d(-t)} \;=\; H\,|\psi>$$

Leaving the Schrödinger equation invariant.

So what the CPT does is depicted in Fig. 3.12.

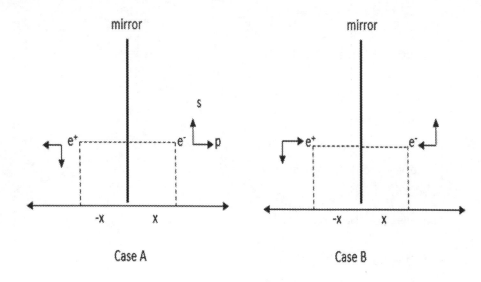

Case A   Case B

Fig. 3.12

Case A: We have an electron at a distance x from the
origin, spin up, and moving to the right. Its mirror
counterpart is a positron, spin down, moving to the left.
CPT means that an equation in QFT describing the right-
handed side of Fig 3.12, Case A, will also describe the
situation in the left-handed side. So what we are looking at
is NOT a situation in which time is reversed in the sense of
taking a trip in the past but looking at a situation in the
RHS of a mirror and its counterpart in the LHS being
consistent with QFT as long as we replace the particle with
its anti-particle of opposite charge and spin, moving in the
opposite direction. Notice in this setup, time is a
parameter, as it should be. Strictly speaking, time reversal
in this context refers to the flow of charge: the particle
moves in the opposite direction of its anti-particle, which
has the opposite charge.

Now suppose that a movie is taken. How do we differentiate if the movie was shown in reverse since there are no collisions and the 3rd law cannot help us? What we would see is Case A and Case B (movie in reverse) as totally different cases in which the actions are reversed. We would not know in which direction time is flowing but we would be able to say that in one case the movie is being played normally, and that in the other case, it is played in reverse, that is, the flow of time is no longer necessary to describe what is seen, but only observing the action (motion) is. And that is in line with what we have claimed in chapter 1, that motion is fundamental, and not time.

3.14 Conclusion

In this chapter we introduced a new law of kinematics. Stated again: given an inertial frame, it will always be the case that the higher kinetic energy particle will lose energy in an elastic collision, and the lower kinetic energy particle will gain energy. This is based on the fundamental notion that a free particle can never have negative kinetic energy. This law gives new perspectives to the 2nd law of thermodynamics which is reformulated as: when two systems at different temperatures are place in contact with each other, the average kinetic energy of the higher kinetic energetic particles will decrease to the limiting point called the thermal equilibrium, while the average kinetic energy of the lower kinetic energetic particles will increase to the thermal equilibrium. This new law of kinematics also explains that the decay of particles is possible only because a source is available through

Einstein's $E = mc^2$, to give each particle after decay positive kinetic energy. It also provides an explanation why the zero-point energy is a necessary condition for the possibility of particles to pop out of the vacuum.

Postscript: Note that the quotation at the beginning of this chapter is often used to promote the debate between two crucial overarching organizing principles of nature: emergence and reductionism. As many would claim that entropy and the whole branch of thermodynamics are a model concept for the side of those who view the universe in terms of emergent laws, the 3rd law of kinematics seems to counter this prevailing attitude as it gives a reductionist answer to why the entropy law does exist. One analogy can be taken from chemistry: the different types of chemical bonding – ionic, covalent, polar and hydrogen – are basically emergent phenomena due to a residual effect of the electromagnetic force, which is a fundamental force of nature. Similarly, entropy is an emergent phenomenon of the 3rd law of kinematics, which it also a fundamental law of nature.

3.15 (Optional) The Case Against Black Holes

The Schwarzschild radius is defined as:

$r_{\text{Schwarzschild}} = \dfrac{2Gm}{c^2}$, where $G = 6.674 \times 10^{-11}$ m^{3}kg^{-1}s^{-2}.

The sun with a radius of 6.975×10^8 m has a mass of 1.989×10^{30} kg, giving a Schwarzschild radius of 2.95×10^3 m.

For the earth, its radius is 6.378×10^6 m, and its mass is 5.972×10^{24} kg, giving a Schwarzschild radius of 8.87×10^{-3} m.

Note: in both cases, $r_{\text{Schwarzschild}} \ll r_{\text{object}}$.

Assume the sun is collapsing. In this scenario, the sun will eventually collapse to the mass size of the earth. That means that its Schwarzschild radius decreases from $r^{sun}_{\text{Schwarzschild}} = 2.95 \times 10^3$ m to $r^{earth}_{\text{Schwarzschild}} = 28.87 \times 10^{-3}$ m. This is a general feature for collapsing stars: as they are decreasing in mass, their Schwarzschild radius is also decreasing. Therefore we must conclude that

$$m_{star} \to 0 \ then \ r_{\text{Schwarzschild}} \to 0$$

In other words, the collapsing star **never** becomes a black hole. In the original paper of Oppenheimer and Snyder [14], it was assumed that the collapsing star did NOT radiate any energy - one could theoretical squeeze a star to its Schwarzschild radius with no loss of mass. However such cases do not exist in the real world.

"Learn the **rules** like a pro, so you can **break** them like an artist."

— Pablo Picasso

Chapter 4

Quantum Mechanics and its interpretation

(History is still haunting us)[16]

Some preliminary remarks about cause and determinism.

Case A: flipping a fair coin

We know that the outcome of flipping a fair coin is 50% heads, 50% tails. Should I flip that coin, say 1,000 times, and find that 60% of the outcomes were heads, 40% tails, this would demand looking for some causes. Perhaps it isn't a fair coin, or there are other factors influencing the outcome such as some unknown magnetic force, whatever, the point is that such a result is in need of an explanation. Now, why is it random in the first place? To be able to predict the outcome on every toss, I would need to know exactly what force was applied on each toss. What I mean by exactly is that I should be able to measure each force to an infinite number of decimal digits. Likewise I would have to know exactly where that force was applied. I would also have to know every other factor — wind, temperature, pressure, any fields such as gravitational, electromagnetic, and so on. To accomplish this requires a considerable challenge from a practical point of view.

Case B: the decay of an atom

Consider the simple case of the decay of a neutron into a proton, an electron and an anti-neutrino, which takes approximately 15 minutes. Now we don't know how the internal clock of a neutron is set to know exactly when a neutron will decay. You can set up a thought experiment in which you can fabricate neutrons, say 1,000 neutrons. But unless you can fabricate all of them at the same time, when examining that sample of 1000 neutrons, they would still decay at random. To offset that you would need to know the theory behind that determines how this internal clock is set. So here we have not only a practical challenge but also a theoretical one.

Now note that in both of these cases, regardless that we do not know how to do it (either on a practical or theoretical level) we still have causes (factors) that underlie the randomness.

Case C: Quantum mechanics

Here's a popular misconception: "QM says everything is uncertain until measured."

As you might already know QM is a probability theory. However in QM, we have a situation totally different from cases A and B. Here the observer in her endeavor to measure a certain property of a quantum system, ends up destroying the original setting of that property. Furthermore, there is no escape from that.

Suppose our observer wants to measure the spin of an electron. The only way we know how to do this is to pass the electron into a magnetic field. She picks an electron at random. We can fairly guess its spin points in any direction. But then when that electron passes through the magnetic field it will flip such that its component along the direction of the magnetic field will be up or down. We do get to know one out three components, but only after the measurement. Whatever its spin was before the measurement remains unknown. So regardless if we can meet both the practical and theoretical challenges, we still are out of luck. It's not that QM says that everything is uncertain until measured but rather that when a measurement is taken, the original system is modified due to an interaction between the system and the observer. We nevertheless get partial knowledge but about a system that has been changed in the measuring process.

Concerning the debate on the interpretation of Quantum Mechanics (QM), historically, Niels Bohr won the debate in establishing what has come to be known as the Copenhagen Interpretation (CI). But Einstein was right in his objection, pointing out that Bohr's interpretation leads logically to a spooky action at a distance. Unfortunately, Einstein's own remedy to this situation – "hidden parameters" – also failed. This controversy was compounded by Bell's theorem, which did not resolve the hidden parameter question, nor the spooky action at a distance. It basically reaffirmed the probabilistic nature of QM as suggested by Born. However Bell's theorem also open the door to another debate involving the collapse of the wave function and non-locality. Now QM is neither mysterious nor weird. You would do yourself a great favor if you would take that idea out of your system. A word of caution: the "wave function" should be used keeping in mind that it does not refer to a real, physical wave but like the Lagrangian, it is a function, most particularly, a function that satisfies the Schrödinger equation, one of the fundamental axioms of QM (see at the end of this chapter). In that sense, it is a mathematical object whose primary role is to calculate probabilities.

In this chapter we will illustrate the salient points of QM and hopefully bring logical clarity in that debate.

4.1 Probabilities

Let's clarify immediately one point: quantum states are not real. They are a mathematical device to calculate probabilities. And here's another secret: probabilities are in themselves not real (not to confuse with representing probabilities with real numbers, which is another issue) but as you should have guessed, they are a mental construct as any math concept is. Consider the probability of rolling a double "6" from two six-sided dice is 1/36. But to get exactly that number, you would need to roll that pair of dice an infinite number of times. The reality is one could roll a pair of "6" on the first roll or after 100 rolls. It's only after a very large number of rolls that we would observe this calculated probability to be 1/36.

QM allows the calculation of probabilities of such phenomena as the energy emitted/absorbed by an electron as it moves through different energy levels in the hydrogen atom, the spin of an electron as it moves into a magnetic field, where a photon will land on a screen after passing through a double slit, and so many other phenomena. Doing the required experiments, calculating the probabilities from many observations, then comparing these with the calculated probabilities from the theory will tell us if yes or no we have a good theory. The verdict is out: QM is an excellent theory.

At the heart of QM is the question: what is the trajectory of a particle – "Matter moving through space"? And the answer is not what many interpretations of the past have alluded to: that suddenly we realize the universe is no longer deterministic. It was never a question as to whether or not the universe is deterministic. Consider Fig. 1.5 of chapter 1, reproduced below as Fig 4.1,

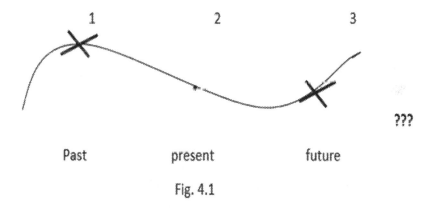

Fig. 4.1

What happens next after position 2 (our present) are a gazillion of possibilities. What is meant by the single principle, "Matter moving through space" is what is common to every phenomenon. But one cannot reproduce every phenomenon from that single principle.

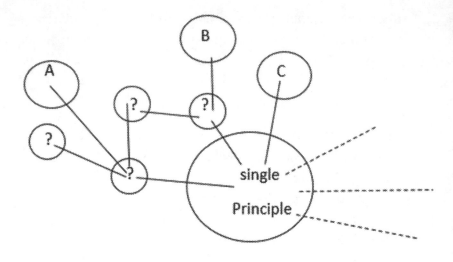

Fig. 4.2

We can go from any of the phenomena A, B, C... to the single principle. But to go back from the single principle to these phenomena A, B, C... would require the knowledge of many other principles, some presently unknown.

In QM, the problem of our investigating is put to the task. The trajectory of a particle at microscopic scale is not completely available to our investigation. And it's not that the particle has acquired a mind of its own, nor that it lives simultaneously in many different universes or in many different quantum states.

Immediately we have two problems to deal with in QM:

1) The electron is never at rest (the position problem);

2) To know its momentum, you need some sort of collision or interaction (the momentum problem).

3) As a result, making a measurement of some sort will alter the configuration of the initial system.

4.2 The Heisenberg Uncertainty Principle

A re-interpretation of the HUP is in order.

Here's a thought experiment. Suppose you were God and you could grab an electron and deposit at a certain position. As God, you've just violated the HUP – but that's okay, God can do that. We could depict this as in Fig. 4.3.

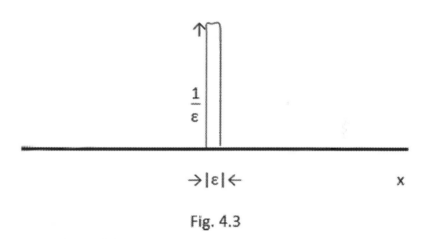

Fig. 4.3

But as soon as you (or God) release(s) the electron, it would look like Fig. 4.4.

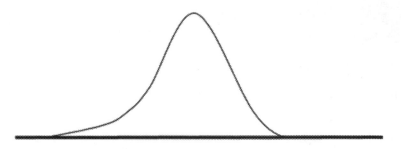

Fig 4.4

After a certain time, the position of the electron has spread out. The question is: what does that tell us? It looks like the particle is doing some kind of motion, some jiggling. It means that for microscopic particles, they are never at rest. In classical physics, you can have objects at rest. The walls in your room are at rest with you. But in QM, no object is at rest. And that's a fundamental difference with classical physics. In other words, at quantum scales you can never find a rest frame for an observer outside the frame of the particle.

What else does the HUP tells us?

How do we measure the velocity of a car? I see the car because an enormous number of photons are hitting the car in all directions, and some of them will reach my eyes. I can then note where it is at a given time, call that x_1 and t_1. At a later time, I observe the car again but at x_2 and t_2. I can get a whole set of these points, plot it, get the velocity, determine if it is in uniform motion or if it is accelerating or decelerating, etc.

But consider the case of an electron.

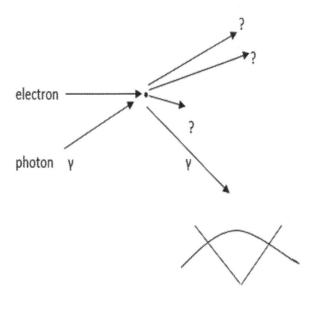

Fig 4.5

So it goes for an electron, to find out anything about it, the idea is to shoot a whole bunch of photons, in this case short-wave or high-energy photons due to the smallness of the electron. We get lucky as one of those photons hits the electron, and with very much luck, it bounces in the right way to reach our eyes. This is what the photon would be telling us if it could speak, "Sir, that electron is right there," call that position X, even though X is really a smeared area as our electron was jiggling around when it was hit, "but then guess what Sir, I've also thrown it off its position, and I haven't a clue in what direction it's going." Hitting the electron with a photon would be like hitting the car in the previous case with a missile. It would be unlikely that the car would have continued along its path. It would follow some undetermined path as in Fig. 4.5. Hitting the electron with a second photon to get another position and time, x_2 and t_2, would be a difficult task. In other words, the electron's path after the collision with the photon becomes unpredictable. It can still be dealt with a probability theory, which is what QM is, by placing a series of detectors surrounding the location of impact.

This is the second thing the HUP is saying: if the position is known with zero uncertainty, then its momentum is unknown. And likewise, if the momentum would be known with absolute certainty, then its position would also be unknown. This is characterized as,

$$\Delta\sigma_x \, \Delta\sigma_p \; \geq \; \hbar/2 \qquad\qquad 4.1$$

Where σ is the standard deviation. Note that in the case of the car (a classical system), we need not to worry that the photons will disturb the trajectory of the car.

The third thing that the HUP says is that if you make a measurement, the very act of making the measurement will alter the system. In our case, we had an electron jiggling about the position X, and now, it's jiggling somewhere else.

To resume, for a quantum system:

(1) A particle is never at rest.
(2) There is an uncertainty in measuring the position and momentum at a given time as indicated by equation 4.1.
(3) A measurement on the quantum system alters the system in some unpredictable way.

The net result is that we get partial knowledge of the quantum system, and we have to make do with a probability theory to deal with that reality. This also shows another assumption that Classical physics was hiding – the other assumption was discussed in section 3.12.3 – and that is, we could observe a system without affecting it. QM uncovered this assumption and rectified it.

4.3 Incompatible Observables – Conjugate Pairs

Consider a number of plane waves moving towards a slit as in Fig. 4.6.

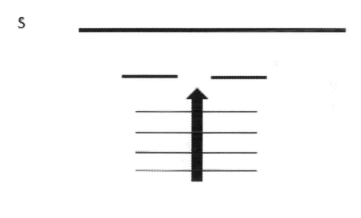

Fig 4.6

As they go through the open slit, they start to bend and will hit the screen S, leaving on it a series of white and dark fringes (Fig. 4.7a).

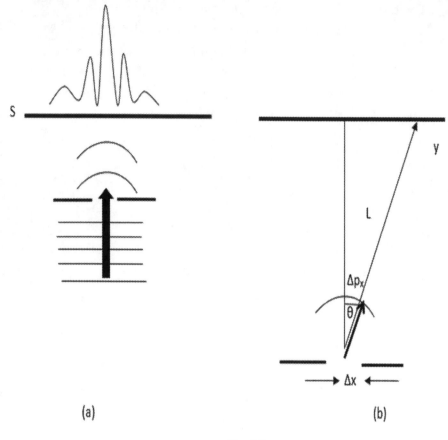

S

(a)   (b)

Fig. 4.7

Consider a point one wavelength away from the slit (Fig 4.7b), which will travel in a straight line and hit the screen at point y and note that the wavelength λ is << L.

$$sin\ \theta \approx \theta = y/L \qquad\qquad 4.2$$

Also, when that point was entering the slit, it had only momentum along L. But now it's moving at an angle θ and has developed a momentum along the x-direction, Δp_x.

$$\Delta p_x \approx p\,\theta \qquad\qquad 4.3$$

Moreover, the point on the wave is one wavelength λ, and $\Delta x/2$ away from the line of motion,

$$\lambda \approx (\Delta x/2)\,\theta \qquad\qquad 4.4$$

Now, historically, de Broglie had proposed that every particle is associated with a wavelength such that,

$$p = h/\lambda \qquad\qquad 4.5$$

Where h is Planck's constant. Combining equations 4.3, 4.4 and 4.5, we get

$$\Delta x\,\Delta p_x \approx 2h > h \qquad\qquad 4.6$$

This is also the HUP, expressed in terms of the uncertainty in the position Δx, and the uncertainty in the momentum Δp. If we want to reduce the uncertainty in the position by passing the waves through a smaller slit, then the bending of the waves will be more pronounced, and so the uncertainty in the momentum will be larger. And vice versa, to decrease the uncertainty in the momentum requires less bending, and to accomplish that the slit must be wider. We say that the position and the momentum are incompatible observables. These are often called conjugate pairs.

4.4 Quantum States

Let us be clear: a particle does not exist in a state of superposition – a silly idea still persisting up to the present. The superposition principle is a valid concept in so far as it postulates the possible states that a particle can have. Also it is very important to be reminded that those possible states before a measurement takes place are NOT open to investigation. When a measurement of a certain property of the particle does take place, because an interaction had to enter into the fray, that property has been altered and what was its state prior to the measurement no longer exists. However, once a measurement has been taken, the beauty in some situations – for instance, in the case of measuring the spin of an electron – is that we can now repeat this measurement over and over and get the same result.

Another confusion in Quantum Mechanics is the result from not being able to differentiate between the real world and the Hilbert Space. Vectors in real space – like velocities, accelerations, forces, etc. – are objects one can actually measure in the real world. On the other hand, quantum states are represented by vectors (more precisely by rays) in a Hilbert space, but these are NOT subjects of measurement. What we measure for a quantum system are probabilities, and those vectors in that Hilbert space are useful mathematical tools to calculate those probabilities. Suppose we have a beam of electrons flowing from right to left:

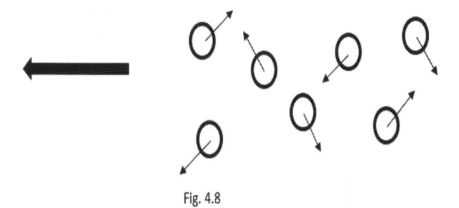

Fig. 4.8

Notice this is a thought experiment as we really don't know in what direction the spin of each individual electron is pointing. We can safely say that these directions are at random. Yes, Einstein was corrected in this particular case: "... there exists a physical reality independent of substantiation and perception" [17]. But we need to make observations if we want to understand the underlying principles that govern the universe. And we can't help ourselves but to interfere when we do that. We can also say that these particles are not entangled (more to say about entanglement later on).

Now physicists are interested in measuring these spins.

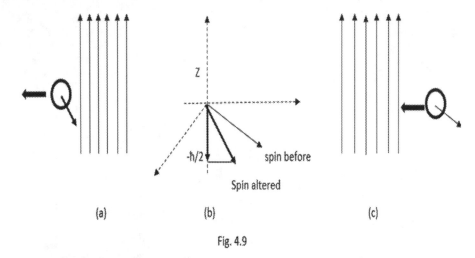

spin before

Spin altered

(a)   (b)   (c)

Fig. 4.9

So what is needed is some kind of apparatus, and the good news is that there exists one – a magnetic field. Trouble is that these electrons, with their spin, are tiny magnets, and we know that magnets placed in a magnetic field will align (or anti-align) with the magnetic field. Suppose a magnetic field is placed along a certain direction, say the z-axis. Now let's look at one electron as it approaches the magnetic field (Fig. 4.9c).

When that electron penetrates the magnetic field, it will align its spin such that its z-component will yield the value of $-\hbar/2$ along the z-axis, a spin down, which can be represented as in Fig 4.9b.

Note that after passing the magnetic field, the electron's total spin has been altered.

On the whole, 50% of the electrons will align with the magnetic field (spin $=+\hbar/2$, or up), and 50% will anti-align (spin $= -\hbar/2$, or down).

Comments

(i) Before the measurement, the spin of an electron can be in any direction. Passing the electron through the magnetic field forces the electron to change its spin orientation such that it either aligns or anti-aligns with its z-component to be $\pm \hbar/2$. This is what distinguishes quantum physics from classical physics: the act of measuring a quantity will disturb the system.

(ii) The other components of the spin are indeterminate: if I were to pass these electrons into another magnetic field, say aligned with the x-axis, again it will be found that 50% of the electrons will align with the magnetic field (spin = $+\hbar/2$), and 50% will anti-align (spin = $-\hbar/2$), this time along the x-axis. On the other hand the spin along the z-axis is no longer known for these particles.

(iii) One way to mathematically represent this quantum system (read, the wave function) is this:

$$|\psi> = (1/2)^{\frac{1}{2}} (|\uparrow> - |\downarrow>) \qquad 4.7$$

Now as it was already mentioned, this is called a superposition of two quantum states, the up and down states. Note that if we want to calculate the probability that the electron has a spin up, we take the product of the vector $|\uparrow>$ with the wave function $|\psi>$, and square that.

$$P = |<\uparrow|\psi>|^2 \qquad 4.8$$
$$= 1/2 [<\uparrow|(|\uparrow> - |\downarrow>)]^2$$
$$= 1/2 [<\uparrow|\uparrow> - <\uparrow|\downarrow>]^2$$

Using the orthogonality condition, which is a fundamental property of a Hilbert space,

$$< \uparrow \mid \uparrow > \; = \; 1 \; and \; < \uparrow \mid \downarrow > = \; 0$$

We get,

$$P = 1/2, or \; 50\%, \qquad\qquad 4.9$$

Which is what is observed in the lab.

(4) After the electron has passed through the magnetic field, if passing again through the same magnetic field, the result will be the same. This is the notion of preparing a particle in a given quantum state.

(5) Now here comes the real crunch. Writing $\mid \psi > = (1/2)^{\frac{1}{2}}$ ($\mid \uparrow > - \mid \downarrow >$) is called a superposition but it's not meant to mean that the electron "lives" simultaneously in two states and can't make up its "mind" in which one it wants to live. Equation 4.7 does not mean that the electrons live in two different states, but rather that it indicates the possible outcomes should a measurement be taken. The superposition of these states is not about how weird QM is but about the mathematical structure of a Hilbert space.

To be kept in mind is that those states do not represent ordinary vectors of real objects - like velocities, acceleration, forces, which was mentioned above. If it were the case, then since these two vectors are equal in magnitude and opposite in direction I would be able to claim, $\mid \uparrow > = (-1) \mid \downarrow >$. And the orthogonality condition would no longer hold, and P would not equal to 50% - actually it would turn out to be 100%!!! What needs to be reminded is that the two vectors, $\mid \uparrow >$ and $\mid \downarrow >$

represent <u>possible states before the measurement</u> takes place. And the beauty of it all is that they form a complete set of orthogonal unit vectors, in an abstract space called the Hilbert space, which provides a powerful method of calculating probabilities.

4.5 A Second Look at the Two-Slit Experiment

By the principle of linear superposition, two states will evolve as,

$$|A> + |B> \rightarrow |A'> + |B'>$$

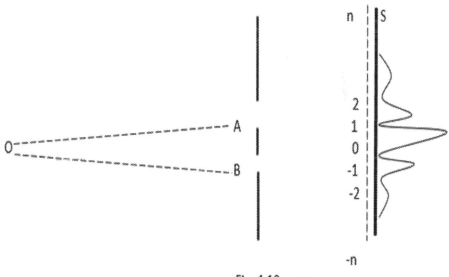

Fig. 4.10

We label the position from which the electrons pass through as such: where they leave is O; the slits are labelled A and B; and the screen, S (see fig. 4.10). The state representing the electron passing through slit A is denoted by $| A >$ and going through the second slit B as $| B >$. When the electrons leave position O, from the symmetry of the setup, we can say that they arrive at position A and B with equal probability. So we write,

$$| 0 > \rightarrow \quad | A > \ + | B >$$

After an electron has arrived at A or B, what happens when it hits the screen? Experiments show that they can land on any of the points on the screen, so we write,

$$| A > \ \rightarrow \ \Sigma_n \Psi_n | n > ; | B > \ \rightarrow \ \Sigma_n \varphi_n | n >$$

Where $| n >$ forms a complete set of orthonormal vectors. The whole process can be described as,

$$| 0 > \rightarrow \quad | A > \ + | B >$$

$$\rightarrow \Sigma_n (\Psi_n \ + \ \Phi_n) | n > \ \equiv \ | \chi_n >$$

The probability that an electron will arrive at the mth point on the screen is the square of the amplitude (equation 4.8),

$$P_m = \ | < \chi_n | m > |^2 = < \chi_n | m > < m | \chi_n >$$

With the amplitude $< \chi_n | m >$ being the complex conjugate of $< m | \chi_n >$

$$P_m = \sum_n \sum_{n'} (\Psi_{n'}^* + \Phi_{n'}^*) < n' | m > < m | n >$$
$$X (\Psi_n + \Phi_n)$$

Using the orthogonality condition,

$$< n' | m > = \delta_{n'm} \; ; \; < m | n > = \delta_{nm}$$

$$P_m = (\Psi_m^* + \Phi_m^*)(\Psi_m + \Phi_m)$$

$$= \Psi_m^* \Psi_m + \Phi_m^* \Phi_m + \Psi_m^* \Phi_m + \Phi_m^* \Psi_m \qquad 4.10$$

The first term $\Psi_m^* \Psi_m$ represents the probability if only the first slit was open. Similarly, the second term $\Phi_m^* \Phi_m$ represents the probability if only the second slit was open. Classically, we should get the sum of these two terms if both slits were open. But we do not observe that. The interesting aspect of this result from quantum physics is that we get two extra terms, $\Psi_m^* \Phi_m$ and $\Phi_m^* \Psi_m$, that correctly explains the interference pattern of the double-slit experiment. Another major difference between classical physics and quantum physics is that in the first, probabilities are added, while in the second, the amplitudes are added first and then we square the amplitudes to get the probabilities.

4.6 The Act of Measuring and Entanglement

Suppose we want to know through which slit the electron has passed. This can be done by inserting a detector at position A. Furthermore, we prepare the electron at position 0 with a down spin. When it passes through A, its spin is flipped to an up spin, and when it passes through position B, nothing happens to the electron and it remains with a spin down (see fig. 4.11).

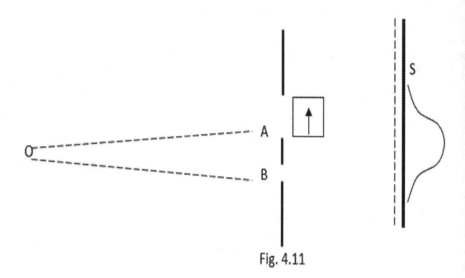

Fig. 4.11

We now need two labels for the states: one for position, and the other for the spin. We describe the process as,

$$| 0, d > \; \rightarrow \quad | A, u > \; + \; | B, d >$$

$$\rightarrow \; \sum_n \Psi_n \, | n, u > \; + \sum_n \Phi_n \, | n, d >$$

Due to the presence of the detector, the electrons are entangled through their spins: one is up, the other is down. Entangled states arise from a particular situation with specific conditions. In other words, they are not arbitrary states. Before we explore more on this, it is suffice to say for now that entanglement means that if we know a certain property of one particle, we also know the property of a second particle. But more specifically, entanglement is often the result of a conservation law, be it the conservation law of linear momentum or angular momentum, without that, one could not possibly write down the proper wave function for these cases. Entanglement is just another way of reiterating conservation laws. Note that entanglement though important in its own right is not a fundamental concept. Underlying it is some conservation law – momentum or spin in most cases, and that is why the particles are strongly correlated.

In the aforementioned case, we know that if the spin at A is up after passing through the detector, we also know that if it passes at B, it is spin down. Again to calculate the probability of finding the electron at the mth position, we square the amplitudes, (equation 4.8).

$$P_m = (\Psi_m^* < m, u \,|\, + \, \Phi_m^* < m, d \,|\,)$$

$$X \, (\Psi_m \,|\, m, u > \, + \, \Phi_m |\, m, d >)$$

$$= \, \Psi_m^* \Psi_m < m, u \,|\, m, u > \, + \, \Phi_m^* \Phi m < m, d \,|\, m, d >$$

$$+ \, \Psi_m^* \Phi_m < m, u \,|\, m, d > \, + \, \Phi_m^* \Psi_m < m, d \,|\, m, u >$$

Note that the up and down vector states, because the electrons have opposite spins, are now orthogonal to each other.

$$< m, u \,|\, m, d > \, = \, 0 \, = \, < m, d \,|\, m, u >$$

$$< m, u \mid m, u > = 1 = < m, d \mid m, d >$$

Therefore,

$$P_m = \Psi_m^* \, \Psi_m + \Phi_m^* \, \Phi_m \qquad 4.11$$

This result is completely different from equation 4.10. We see now that the very act of detecting the spin of one electron, that is, making some sort of measurement, alters the interference pattern. Again, this is another markedly difference between a classical system, in which we can always make a measurement without disturbing it, and a quantum system, in which a measurement entails disturbing the system and getting a different result.

4.7 Bell's Theorem Revisited

As it was mentioned, Bell's theorem has played an important role in the interpretation of Quantum Mechanics. First we will prove the theorem, then we will look at its implications.

In Bell's theorem [18], we make two assumptions in the proof. These are:

A. Logic is valid.

B. A body either has a property A or doesn't have property A.

This is important to understand. The property in question is not necessarily non-locality. It can be anything that a particle possesses and can be measured. Consider the set of all measurements, for which A, B and C are any three measurements, and are independent properties. Examples: A is up or down, B is head or tail, C is red or green, etc. Secondly, the theorem is not about hidden parameters but whether an object has a property or does not have it. Making it about non-local hidden parameters is to doubly compounding the error in misinterpreting Bell's theorem.

<u>Derivation of Bell's inequality</u>

Definition: if an object has property A, we denote that as A+; if not, we denote it by A−:

$$N(A+,B-) = N(A+,B-,C+) + N(A+,B-,C-) \quad 4.12$$

This is true since an object must have the property C or does not have it, which is what the RHS says.

We assert that since all three terms in equation 4.12 are positive, then

$$N(A+,B-) \geq N(A+,B-,C-) \qquad 4.13$$

That is, $N(A+,B-)$ cannot be smaller than zero.

Also,

$$N(B+,C-) = N(A+,B+,C-) + N(A-,B+,C-) \quad 4.14$$

Similar reasoning as above: an object must have the property A or does not have it.

Therefore,

$$N(B+,C-) \geq N(A+,B+,C-) \qquad 4.15$$

Similar reasoning as in equation 4.13, that is, $N(B+,C-)$ cannot be smaller than zero.

Adding inequalities 4.13 and 4.15,

$$N(A+, B-) + N(B+, C-) \geq$$

$$N(A+, B-, C-) + N(A+, B+, C-) \quad 4.16$$

But the RHS of 4.16 gives:

$$N(A+, B-, C-) + N(A+, B+, C-) = N(A+, C-) \quad 4.17$$

That is, an object must have the property B or does not have it. Substituting 4.17 into 4.16, we get,

$$N(A+, B-) + N(B+, C-) \geq N(A+, C-) \quad 4.18$$

And that completes the proof.

To reiterate: a body has a property which can be measured or it does not have that property. For instance, looking at the earth at a distance, one can observe its spin around an axis and that can be measured. On the other hand, looking at the moon, we observe it has no spin. So either a body has a spin (the earth) or it doesn't have a spin (the moon). On the other hand, the electron has a spin, but only one component can be measured. The other two components remain indeterminate once one of the components is measured as it was discussed in fig. 4.8 and 4.9. It is in that mind frame that we must interpret Bell's theorem. That is, the theorem does not deal with incompatible observables but instead addresses the case of an object having a certain property or not. In Bell's theorem, we are dealing with three properties (A, B, C), each property being independent of the other two. It goes without saying that a classical system will not violate Bell's theorem, while a quantum system will. Why? Because in deriving the theorem, nowhere the act of observation was taken into the consideration, which in the case of quantum system, alters the system. In Alain Aspect experiment [19], Bell's inequality theorem was tested by measuring the polarization of photons along three different axes. The inequality was violated as the system understudied was a quantum system. To attribute this violation to non-locality is a major blunder. The violation is strictly due to the HUP: along the three axes, the components of the spin are incompatible observables, while Bell's theorem only applies to a classical system, whose properties A, B, C are independent of each other. So applying the theorem to a quantum system will inevitably result into a violation.

4.8 The EPR Revisited

At the 1927 Solvay Conference, the disagreement between Einstein and Bohr first surfaced and began the debate that has lasted ever since. The first disagreement centered on the notion of a wave collapse. Give Einstein one point (+1) being on the right side. He correctly deduced that the collapse of a real wave would mean the existence of a spooky action at a distance.

The EPR [20] that came subsequently (1935) proposed that there were hidden parameters to explain what Einstein thought was the unexplainable. Give Einstein a (- 1) point. It's a tie as far as Einstein is concerned.

Here's your typical argument that has come through the decades since this disagreement started.

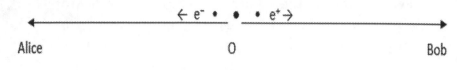

$\leftarrow e^- \bullet \quad \bullet \quad \bullet\, e^+ \rightarrow$

Alice   O   Bob

Fig. 4.12

A particle at O decays and sends two particles: an electron e^- towards Alice, and a positron e^+ towards Bob. Both Alice and Bob are space-like separated.

Each particle flies off in opposite direction with opposite momentum (conservation of momentum) and opposite spin (conservation of angular momentum). And because of this relationship through a conservation law, these two particles are entangled.

<u>Case A</u>

Alice is going to measure the spin of her particles with a magnetic field along the z-axis, likewise for Bob. So both are performing an experiment depicted in Fig 4.8-4.9. In each case, consider one particle at a time. As the electron (positron) approaches the magnetic field, the orientation of the spin with the magnetic field is totally unknown to our two observers. Now the particle goes through the magnetic field. There are only two possibilities for each particle,

(1A) Alice measures a spin up and Bob measures a spin down.

(2A) Alice measures a spin down and Bob measures a spin up.

There are NO other alternatives.

There is no mystery here, there is no spooky action at a distance, there is no weirdness, there is no communication traveling faster than light. There are only two possibilities, and this is what will be observed, which is explained entirely by the conservation laws.

On the whole, Alice will measure 50% of all the electrons coming her way with spin up, and 50% with spin down. Bob will measure similar results for his positrons.

<u>Case B</u>

Alice is going to measure the spin of her particles with a magnetic field along the z-axis, but this time, Bob will measure his particles along a different axis, say the x-axis.

The situation doesn't change in regard to the particle approaching the magnetic field: the orientation of the particle's spin is still unknown to both Alice and Bob.

Consider one particle at a time.

(1B) Alice measures the first particle with a spin up along the z-axis.

(2B) Bob measures his first particle with a spin up along the x-axis.

Can Bob conclude that he also knows that his particle has a spin down along the z-axis, since Alice measured her particle with a spin up along that axis?

No, he doesn't know. His experiment is different than Alice's as his particle's orientation was forced along the x-axis, by an amount that is unknown to him. And Alice's particle was forced to align along the z-axis by also an unknown quantity. The only conclusion that Bob can make is what he measured: a spin up along the x-axis. Secondly the components of the spin of his particle along the y and z axis remains unknown to him, just as Alice doesn't know the x and y components of her particle.

As in case A, Alice will measure 50% of all the electrons coming her way with spin up, and 50% with spin down, but keep in mind, she has only measured the spin along the z-axis. She has no knowledge of the other components of the spin of her particles – the x and y components.

Likewise Bob will also measure 50% of all the electrons coming his way with spin up, and 50% with spin down, but keep in mind also, he has only measured the spin along the x-axis. He has no knowledge of the other components of the spin of his particles – the y and z components.

Again there is no mystery here, there is no spooky action at a distance, there is no weirdness, there is no communication traveling faster than light.

4.9 Spooky Action at a Distance

How can we explain the confusion that has reigned for more than nine decades? There were mistakes done at different levels:

(1) A misinterpretation of Bell's theorem in which the original intent did not include non-locality, but as a test to see whether or not a particle has a certain property that can be measured.

(2) A misinterpretation of the disagreement between Einstein and Bohr. Einstein's objection to the

collapse of the wave function, if that wave is taken to be a real wave, implied a spooky action at a distance, and Bohr should have listened to that. This is cleared up once we understand that the wave function is a mathematical object like the Lagrangian function or the Hamiltonian function, to name a few examples.

(3) A misinterpretation that the wave function represents a real wave when in actuality it represents the possible states of a quantum system. After the measurement, QM yields a probability distribution. And the actual taking place of a measurement alters irrevocably the initial state of that system.

(4) When Bell's theorem is violated by a quantum system, those violations were then interpreted erroneously as evidence of an instantaneous collapse of the wave function, which was labelled as proof of non-locality.

Those who were carrying the torch for Einstein thought that Bell's theorem confirmed non-locality (when in reality Bell's theorem doesn't really say anything about non-locality nor about hidden parameters) because that also confirmed that Einstein was right (a wave function collapse implied a spooky action at a distance, but the wave function isn't a real wave to begin with) leading to the idea that an instantaneous collapse (nothing can travel faster than the speed of light) makes the universe weird.

Here's the real deal: the collapse of the wave function is a fiction, ditto with the spooky action at a distance.

4.9.1 The Pizza Story

Imagine if you have two friends, one loves pizza and the other hates it. After finding out from your Mom that you are having pizza for dinner you decide to tell your two friends. If they are standing next to each other and you yell "Pizza is for dinner" then one friend will yell "YES!" The other will then yell "NO!" This is what you would expect, right?

What if one of those friends went to the Moon instead of your house and you tried this again? This time only the friend that is with you can hear you say "Pizza is for dinner!" Suppose this friend yells "NO!" (Spin down in the EPR). All we can say about the other friend (on the moon) is that he is silent. He can't hear us, we can't hear him.

So what if I told you I would bet you $100 that I know for "sure" that the friend on the moon would yell "YES!" How do I know? How could the friend on the Moon have heard the dinner choice? How could he know Pizza was said at all? Yet, if there is someone on the moon (another experimenter) to ask him, he surely would have yelled "YES!" In this scenario, I am Alice, the other experimenter is Bob.

Well **IF** those two friends were instead two entangled particles, then we would be able to tell 100% sure that the Moon particle would say "YES!"(Spin up) when asked by Bob.

Is this spooky action at a distance? Definitely NOT. The only reason we can do this is that we have some information prior to the experiment – the two friends love/hate pizza – while in the EPR case (Fig. 4.12), equation 4.7 which can be written as such is due to the conservation of spin.

4.10 Classical Physics, Probability Theory and QM

(i) Vectors: In Classical physics, a car in motion can be observed, and an observer can determine its velocity over the course of a given trajectory. This in term can be expressed as a velocity:

$$v = v_1 i_1 + v_2 i_2 + v_3 i_3$$

Where v_1, v_2, v_3 are components and i_1, i_2, i_3 are unit vectors. See illustration below.

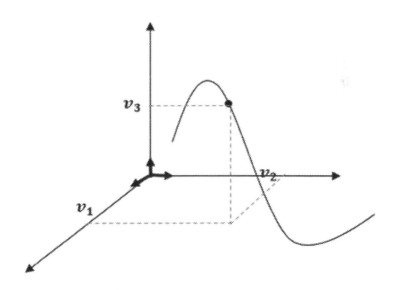

Fig. 4.13

We can express the above equation as:

$$v = \Sigma \, v_j \, | \, i_j >$$

Where j = 1,2,3 in a 3-D world. Note that the unit vectors are orthogonal, that is, $< i_j \, | \, i_k > = \delta_{jk}$

(ii) Probability Theory: Flipping a coin and rolling dices are familiar examples to introduce the concept of probability. For instance, the probability of flipping tails out of a fair coin is $\frac{1}{2}$, or rolling a "5" from a die is $\frac{1}{6}$. This is calculated as:

$$P = \frac{favourable \; outcomes}{total \; number \; of \; possible \; outcomes}$$

Now the sum of all probabilities is 1.

Example 1: Flipping a coin, P_{heads} = ½, P_{tails} = ½, and so the total P_{total} = P_{heads} + P_{tails} = ½ +½ = 1

Example 2: Rolling a die, P_1 = P_2= P_3= P_4= P_5= P_6= 1/6, so the total P_{total} = 1.

(iii) Quantum Mechanics is a hybrid of case (i) and case (ii):

      a) Vector space: An arbitrary quantum state A can be represented as a vector:

$$| A > = \; A_1 \, | \, E_1 > + A_2 \, | \, E_2 > + A_3 \, | \, E_3 > + \cdots$$

Or
$$| A > = \; \Sigma \, A_j \, | \, E_j >$$

Where j = 1,2,3... A noticeable difference with Classical physics is that the dimension of the quantum vector space can be any number between one and infinity. Note that the vectors are also orthogonal, $< E_j \, | \, E_k > = \delta_{jk}$

b) Probability Theory: The probability of finding the particle in a particular state, say E_k, can be calculated as such: First find the amplitude $< E_k \mid A >$. Then take the square of the amplitude:

$$P = \mid < E_k \mid A > \mid^2$$

Substitute the above equation, and taking care of the orthogonal condition, we get:

$$P_k = A_k{}^2$$

To make sure that QM conforms to Probability Theory, we must normalize the amplitudes such that the sum of all probabilities is 1. That is,

$$A_1{}^2 + A_2{}^2 + A_3{}^2 + \cdots = 1$$

Or
$$\sum A_j^2 = 1$$

If the initial arbitrary quantum state $\mid A >$ is not normalized, we can do so by defining:

$$A'_j = \frac{A_j}{\sum A_j^2}$$

And so $\mid A >= \sum A_j \mid E_j >$ becomes $\mid A' >= \sum A'_j \mid E_j >$, which is now normalized.

As it was mentioned at the beginning of this chapter, in Classical physics the act of measuring has no effect on the system being observed. In quantum physics, the act of measuring does alter the system under observation. Because of that very nature that takes place at atomic and smaller scale, Quantum Mechanics is a probability theory, the only choice given by this irrevocable situation. The mystery of Quantum Mechanics is not that Classical physics was tossed aside and we had to start from scratch. In fact Classical physics is all present in QM. So what's so different? Here's a simple way of looking at this: take Classical physics, throw in a few assumptions (app. 6), and you get a probability theory, which is QM. And that's the mystery: how a small number of axioms can yield a theory so different than its source and yet be so close to it, and also gives out predictions that have been overwhelmingly verified by empirical observation.

4.11 The Postulates of Quantum Mechanics

And to close this chapter, here are six assumptions (postulates) of QM, some of which were already mentioned in this chapter:

(i) A quantum state $| \psi >$ can be represented by a vector (more precisely a ray) in a Hilbert space.

(ii) Any arbitrary quantum state $| \psi >$ can be expressed as a sum of orthonormal vectors $| \psi_i >$ in that Hilbert space such that,

$$| \psi > = \Sigma_i A_i | \psi_i > \qquad 4.19$$

Where the A's are complex numbers, and,

$$< \psi_i | \psi_j > = \delta_{ij} \qquad 4.20$$

A specific example are the energy levels, which play a predominant role in establishing the Periodic Table:

$$| \psi > = \Sigma_i E_i | E_i >$$

Where $| E_i >$ are eigenvectors of the Hamiltonian operator, that is,

$$H \mid E_i > = E_i \mid E_i >$$

And the coefficients E_i's are eigenvalues.

(iii) The probability of finding a system in a specific state $| \psi_i >$ is given by the square of the amplitude:

$$P = \mid < \psi \mid \psi_i > \mid^2 \qquad 4.21$$

Unitarity: a consequence of QM being a probability theory is that the theory must be unitary. This is illustrated as follows: Consider an operator U acting on a quantum state.

$$| \psi > = \Sigma_i A_i | \psi_i > \; \rightarrow \; U | \psi > = \Sigma_i A_i \, U | \psi_i >$$

The probability then becomes:

$$P = \mid < \psi \mid \psi_j > \mid^2$$

$$= \mid < \Sigma_i A_i < \psi_i \mid U^\dagger U \mid \psi_j > \mid^2$$

The probability is preserved if and only if $U^\dagger U = 1$

(iv) Observables are represented by Hermitian operators.

For instance, the momentum and Hamiltonian operators of a free particle, which is represented by the wave function $| \psi > = e^{i(px - Et)/\hbar}$, are mapped respectively as:

$$p \rightarrow -i\hbar\frac{\partial}{\partial x} \qquad (-i\hbar\frac{\partial}{\partial x} \,|\,\psi> = p\,|\,\psi>) \qquad 4.22$$

$$H \rightarrow i\hbar\frac{\partial}{\partial t} \qquad (i\hbar\frac{\partial}{\partial t} \,|\,\psi> = E\,|\,\psi>) \qquad 4.23$$

(v) Incompatible observables obey the Heisenberg Uncertainty Principle. For instance, the position x and the momentum p of a particle obeys the commutation relationship: $[\,x,\,p\,] = i\hbar$.

We obtain this result through the following procedure:

(a) In phase space coordinates (x, p) for two arbitrary functions, f(x,p,t) and g(x,p,t), we define the Poisson bracket as

$$\{\,f,g\,\} = \frac{\partial f}{\partial x}\frac{\partial g}{\partial p} - \frac{\partial f}{\partial p}\frac{\partial g}{\partial x}$$

(b) For the position (f = x) and the momentum (g = p), the Poisson bracket becomes,

$$\{\,x,p\,\} = 1$$

(c) Then we impose commutation relation on the Poisson bracket as such,

$$\{\,x,p\,\} = 1 \rightarrow [\,x,p\,] = i\hbar$$

And this yields the Heisenberg Uncertainty Principle.

(vi) The time evolution of a quantum state for a particle is governed by the Schrödinger equation:

$$i\hbar \frac{d|\psi>}{dt} = H|\psi>$$

4.24

Where again H is the Hamiltonian operator.

Classically, the Hamiltonian for a particle moving into a potential V(x) is,

$$H = \frac{p^2}{2m} + V(x)$$

4.25

The quantum version is (using postulate iv),

$$H = -\frac{\hbar^2}{2m}\frac{\partial^2}{\partial x^2} + V(x)$$

4.26

And for such a particle, the Schrödinger equation becomes,

$$i\hbar \frac{d|\psi>}{dt} = \{-\frac{\hbar^2}{2m}\frac{\partial^2|\psi>}{\partial x^2} + V(x)\}|\psi>$$

4.27

We get two equations in the relativistic quantum mechanical version of the Schrödinger equation. The first is the Klein-Gordon equation,

$$(-\partial_t^2 + \nabla^2)\varphi = m^2 \varphi, \quad \text{where } h = c = 1$$

4.28

In Quantum Field Theory (QFT), the K-G equation is interpreted as the equation of a real scalar field, spin-0 particles, which are bosons (integral spins).

The second equation is the Dirac equation:

$$i\gamma^\mu \partial_\mu \psi - m\psi = 0$$

4.29

Where the γ's are the Dirac matrices.

In QFT, the Dirac equation is reinterpreted as describing quantum fields corresponding to spin-1/2 particles, which are fermions (half-integral spins).

Postscript 1

Here's another way one should think of QM:

(a) You write down your wave function of possible states for the system you want to observe.

(b) You include all the info you already know about the system such as conserved quantities (entanglement) into your wave function.

(c) You plug that into the QM machinery. The probability you get is what your observation should confirm.

Postscript 2

What about the wave/particle dual nature of matter? There might be some interest in looking at this debate from a historical perspective. However, when studying waves, Fig. 4.7, absent in that discussion is that water waves are made of water molecules (particles), sound waves are made of air molecules (particles), and so it should be no surprise that light waves are made of particles (photons). The concept that a wave needs a medium to propagate is misleading. What you have in all of these cases is energy being transferred from particles to particles - matter moving through space - the medium is just unnecessary baggage. Take for instance a series of cars waiting at a red light. When it turns green, the first car moves, increasing its distance between itself and the second, then the second car starts moving increasing its distance from the third card, and so on. As this train of car meets the next red light each car will slow down and the distance between the cars will diminish. If one concentrates solely on the spaces between the cars, you will see a wave moving throughout, yet it is the cars that are actually moving.

Postscript 3

A wave function is associated not with a beam of particles but with each particle. In the case of a beam of monochromatic light, every photon is given the same wave function Ψ, and the same probability, so that the beam reproduces the expected classical observation.

Postscript 4

Perhaps the "wave function" should be renamed as it does not describe a real wave – it is a mathematical object. My humble suggestion would be to rename is as the "Schrödinger" function as it is **the** function that is the solution to the "Schrödinger" equation (postulate vi).

Postscript 5

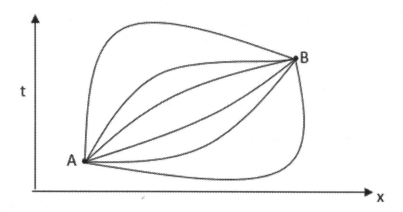

Fig. 4.14

An alternative to the above postulate (vi) is the Feynman's path integral. We assume that a particle will travel from some point A to point B (Fig. 4.14).

We know from previous arguments that if the particle is at point A we also know that it is not possible to know in which direction the particle will proceed in its trajectory (Fig. 4.5). But if we insist that it will go through point B then we must consider all possible paths between points A and B. You shoot an infinite number of photons at an electron at point A. You get an infinite number of paths. There is a subset of paths, also infinite, that goes through B. What Feynman did is to attach to each of those paths a weighted phase factor $e^{iS/\hbar}$, where S is the action (equation 2.39). By summing over all paths – often referred as the "sum of histories" – he was able to derive the Schrödinger equation (equation 4.27) [21].

GR is a theory about a special frame – the free falling frame – in which all the laws of SR are assumed to be valid.

Chapter 5

New Insights into General Relativity

(In GR, Gravity Behaves Like a Fictitious Force)

A pilot at rest on the ground pretends to be in an inertial frame. He takes off in his plane and notices an unusual effect – the Coriolis force. An observer falling freely under gravity pretends he is in an inertial frame. He notices an unusual effect – the bending of light under gravity.

In regard to the fundamental forces of nature, there are basically two categories: (1) except for gravity, all of the other fundamental forces are of the Yukawa type of interaction between particles; (2) forces arising from non-inertial frames - gravity behaves analogously to a Coriolis force, that is, it is a fictitious force. We can look at this from this point of view:

(i) Rotating frame → Coriolis force.

(ii) Presence of matter → gravity.

Of course there is a difference between (i) and (ii): in the first case, we can always switch off the rotation of a frame; on the other hand, we cannot switch off gravity. What Einstein essentially did was to create a special frame – the free falling frame – in which we can pretend that gravity is switched off, and claim that all the laws of SR still apply in particular light interacting with gravity. Noteworthy: Galileo also did something similar in his famous inclined plane experiments, ignoring the force of gravity, and thus developing the first law of kinematics, which we will explore in section 5.4. And so whenever we have to deal with light interacting with gravity, GR is the right theory because it was specifically constructed to deal with that situation.

5.1 General Comments

This brings us to the main focus of this chapter, which is:

(1) That General Relativity (GR) is not a theory of gravity but an extension of Newtonian Gravity (NG) that includes Special Relativity (SR), that is, the speed of light is an invariant and that time is no longer absolute. It is not a coincidence that the effects predicted by GR are about gravity but that doesn't make GR a theory of gravity.

(2) That NG applies to the solar system - it stems from Kepler's laws of the planets orbiting our sun - but does not necessarily apply anywhere else. It is universal in that all matter attracts but the inverse squared law might not be as universal as we think it is. As we know there is a major problem with the velocity of stars in the galaxy – both GR and NG have failed to account for this anomaly.

(3) The irony is that GR was based on the idea that there are no absolute frames of reference in the universe yet it turned out to be a theory invested heavily on the metric tensor which is dependent on the choice of a special coordinate frame.

Historically, it became to be believed that GR was the correct theory of gravity and NG was an approximation to

GR, and all hopes were pinned on GR to give a complete picture of the cosmos. The proposal here is to re-examine this historical consolidation. The prestigious status of GR acquired through the intervening years since its inception has been overstated. We need to develop a complete theory of gravity - one that can explain galactic stellar motion (GSM). Therefore it's preposterous to believe that either GR or NG can explain the cosmos when neither can explain GSM. A theory that can't explain the galaxies is like trying to explain the Periodic Table without QM. There is a need of a better theory than NG or GR if we ever hope to explain the universe – one that will at least explain GSM. A complete classical theory of gravity is missing.

Note that at atomic and sub-atomic scales, energy is quantized in regard to emission and absorption. In gravity, we don't know if that is the case: the detection of gravitational waves as observed by LIGO does not move in that direction. Yes the detector recorded the collapse of two seemingly black holes, but nothing indicates that these energy bursts arrived at our detectors in the form of "gravitons".

In physics, one of the realities that physicists are confronted with is writing theories and laws in terms of mathematical equations. This often requires a choice of coordinates. And since we must make sure that the equations yield the same results for different observers who happen to be in different frames of reference, each observer making perhaps a different choice of coordinate system, then how one translates one equation from one frame of reference to another is of crucial importance.

The genius of Einstein was to realize that gravitational fields were relative in that one can choose a frame in which gravity is removed. For a free falling frame, the gravitational field disappears and the laws of Special Relativity (SR) can be applied as if the free falling observer was in an inertial frame. This led him to the Equivalence Principle (EP) and subsequently to General Relativity (GR). Underlying both SR and GR is that there is no absolute frame of reference in the universe. And at the same time, Einstein's EP extended his SR from inertial frames to a non-inertial frame, the foundation of GR, albeit a very special non-inertial frame. With the use of a freely falling frame as a pivot between inertial frames and non-inertial frames for which a = g, it is that insight which led to Einstein's theory of General Relativity (GR).

5.2 Gravitational Redshift

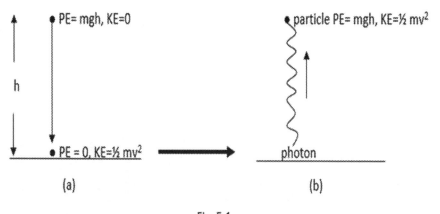

Fig. 5.1

One of many considerations that concerned Einstein was that gravity had to affect light. Why? He imagined this thought experiment. Suppose that gravity does **NOT** affect light. Then one could construct the following scheme:

One could release a particle of mass m at rest at a distance h from the ground (fig. 5.1a). As it falls through the gravitational field, its potential energy (PE) is converted to kinetic energy (KE). Einstein knew that this particle could then be converted to a photon (fig. 5.1b) from $E = mc^2$, and then one could send the photon to climb against gravity, reconverted to a massive particle after it has climbed the same distance h but now with KE=$\frac{1}{2} mv^2 \neq 0$. Comparing (a) and (b) we see that our particle at height h has gained kinetic energy. One could then repeat this process and create energy out of it. Einstein reasoned that the law of conservation of energy demanded that the photon, a massless particle, must lose energy when climbing up against gravity just like any other massive particle. But how, since light always travels at a constant speed c and a photon has no mass? The only way out was to use what he had already used in his seminal paper on the photoelectric effect,

$$E = \hbar\omega.$$

If E has to decrease, then the angular frequency ω must also decrease, or its wavelength increase. This is known as the gravitational redshift. Note that this phenomenon can be explained from the conservation of energy, the conversion of matter to energy back to matter ($E = mc^2$), and light in the form of a particle that can lose energy by extending its wavelength ($E = \hbar\omega$). There is no need of GR.

5.3 General Relativity

The motivation behind GR was to extend the principle of Relativity from inertial frames to non-inertial frames. It is not a theory of gravity "interacting" with light as in QFT, which deals with interactions of the Yukawa type. While SR deals with inertial frames, GR deals with non-inertial ones. So it is misleading to view GR as a theory of gravity, though gravity certainly plays a central role. The main concern for Einstein was to see how light behaves in non-inertial frames. In doing so, he narrowed down to one specific non-inertial frame: the one with a free falling frame.

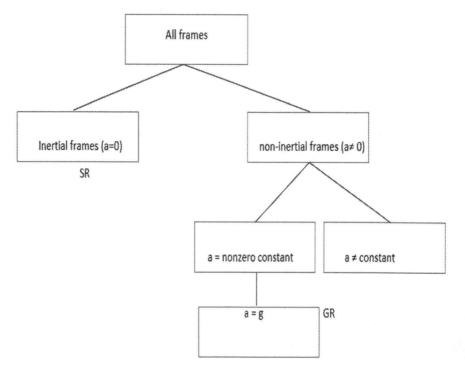

Fig. 5.2

Consider the set of all frames, Fig. 5.2. It contains two subsets: inertial frames (a=0), and non-inertial frames (a≠ 0). SR deals with the former. In the sets of non-inertial frames, there are two subsets: a = a nonzero constant, and a ≠ a constant. In the sets containing a = nonzero constant, there is the subset in which a = g, the acceleration due to gravity. GR deals primarily with this set.

In gist this is what we have:

(i) Observer A in an inertial frame makes a measurement "a".

(ii) Observer B in a different inertial frame make a measurement "b" on the same event.

SR is the machinery that allows Observers A and B to compare their measurement.

With GR, we get the following.

(iii) Observer C in a non-inertial frame but free falling makes a measurement "c", but now we can use SR to relate his measurement to observers A or B.

Let met outline what this last claim is all about.

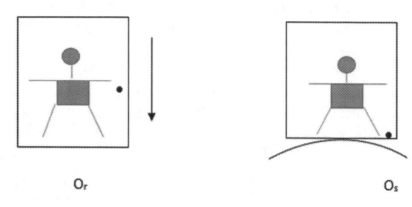

Fig. 5.3

In GR, we construct two special reference frames (Fig. 5.3): the first is a frame in which the observer O_R is freely falling and for whom the laws of physics (SR) are valid – the ball is at rest with the observer O_R; the second is a frame often called the "lab" frame in which the observer O_S is on the ground where the force of gravity is cancelled by the force of the ground preventing O_R from freely falling under gravity. The argument (EP) is that for observer O_r, should he be in a closed room in which he is weightless and therefore floating, and if he is pulled towards one side (the floor with respect to that observer), he wouldn't be able to tell the difference between a force pulling his room or a gravitational field pulling in the opposite direction (Fig. 5.4 below). So Einstein's field equations are formulated for these two special cases – how to transform from one set of frames (O_r) to a second set of frames (O_s).

In a rotating frame, one gets additional features such as the Coriolis force and the centrifugal force which are not present in a rectilinear motion. In GR, the additional features are the correction to Mercury's perihelion and the bending of light. In both cases, these are fictitious forces as they are not due to interactions between particles but due to a transformation of coordinate systems - when gravity is present in the case of GR. From this perspective, GR is not a theory of gravity so much different from Newtonian physics, but rather a theory that takes into considerations the effects of non-inertial frames. What was a revolution initiated by Einstein was the theory of Special Relativity (SR), in which our notion of time was turned upside down. GR contains that revolution in so far as it is an extension of SR from inertial to non-inertial frames.

There is something more to be said about the Equivalence Principle (EP). Einstein arrived as his EP with the following thought experiment. In a room without windows, one would not be able to distinguish between being pulled up by an inertial force F_i (Fig. 5.4a) from being pull down by a gravitational force F_g near a planet (Fig. 5.4b).

But is this true? Are there tidal forces that could show differences?

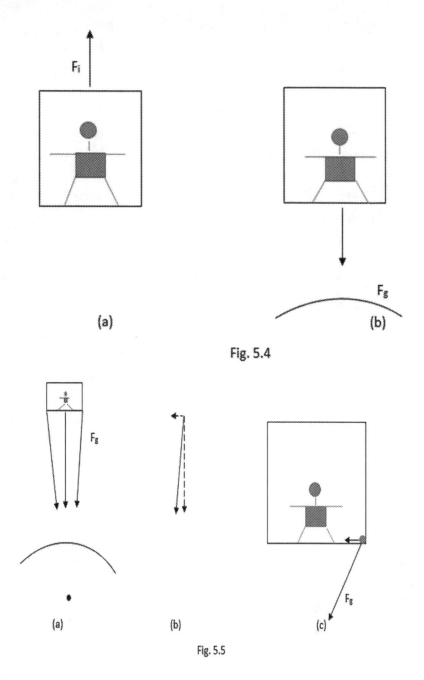

Fig. 5.4

Fig. 5.5

Acting on the frame are tidal forces as illustrated in Fig. 5.5. What the observer would feel besides the downward pull of gravity, is a force perpendicular as illustrated in fig 5.5b. In Fig. 5.5c, a ball placed near a corner experiencing that tidal force would be moving towards the center of the room.

Every point on the right-hand side in that frame would feel this perpendicular force towards the left. Similarly, every point on the left would feel a force towards the right. But what about the Equivalence Principle? Does it predict also this perpendicular force in the case of an inertial force acting on the frame? To examine this question, we need to realize that an inertial force does not exist ex nihilo. What could have created this inertial force?

Consider a collision, Fig. 5.6.

Initially you and the ball are weightless, floating in space, Fig. 5.6a. On the basis of Newton's 3rd law of motion, the frame applies a force on the colliding body, while the colliding body applies on the frame a force of equal magnitude but opposite direction. You will be pulled towards the bottom of the frame (Fig. 5.6b). We also see that after the collision, the floating ball in your frame after experiencing a force that has a perpendicular component, will hit the bottom and accelerate towards the left (Fig. 5.6c).

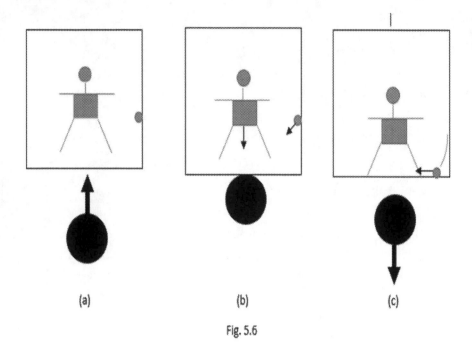

Fig. 5.6

Now, since you are in a room without windows, you could say that:

(a) Either an inertial force acted on you as in Fig 5.4a,
(b) Or you have enter a gravitational field as in Fig. 5.4b.

In both cases you would see the ball accelerating towards the center of the room.

5.4 Straight Line or Curved Line

It used to be believed that in order to move, you needed a force. This was based on every day observations: a carriage had to be pulled by a horse; you needed to row a boat or have a sail nailed to the boat and let the wind push it. This notion dominated the Western world and went all the way to Aristotle, for nearly two thousand years. So the experiment that would overturn this idea is perhaps the most important one, at least as far as science is concerned. Nevertheless, there is a glitch in that experiment that went unsuspected.

Consider how Galileo arrived at the Law of Inertia.

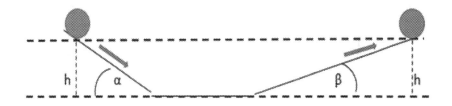

Fig. 5.7

A ball is released from a height h on an inclined plane making an angle α with the ground (fig. 5.7). It would then roll down and climb a second inclined plane making an angle β with the ground. What Galileo observed is that the ball would travel far enough until it would climb a height h where it would come to rest temporarily and then reverse course. By varying the angle of the second plane, he always observed that the ball would climb to a height h before reversing direction. See fig 5.8.

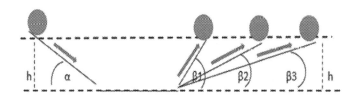

Fig. 5.8

This led him to a thought experiment: what would happen if the second inclined plane were to be removed?

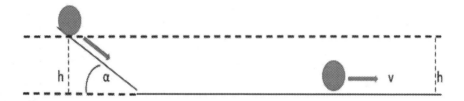

Fig. 5.9

He reasoned that the ball would continue endlessly in a straight line with a velocity v trying to reach the height h, unless other forces would compel it to change its velocity (Fig. 5.9). Now this falls into the realm of an assumption as it cannot be proven in the real world - there are always other forces that will act on the ball, for instance, friction to name one.

On the slope, that force is at an angle, and why the body has a net force along the slope and so it will accelerate down the slope until it reaches a velocity v. Once it reaches the bottom of the slope, these two forces are now equal and opposite, giving a net force of zero, and so the ball will continue to move on with velocity v. But is the ball really going in a straight line?

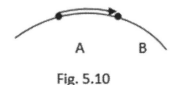

A   B

Fig. 5.10

Had Galileo extended his experiment on a much larger lengthy floor – from Florence to London, for instance - he would have realized that his ball was actually moving along an arc, called a geodesic, as in Fig. 5.10. While Einstein had deliberately switched off gravity in his famous elevator thought experiment (Fig. 5.4), Galileo had in effect unknowingly switched off gravity.

We can now restate the 1st law of kinematics (sec.3.1): unless other forces compel it to change its velocity, an object will continue endlessly with a velocity v, along a path in a curved line. The alternative to that, which would be the modern version, is the ball entered a region of curved space-time. Fortunately, Galileo escaped this conclusion, otherwise his law would have been met with great skepticism.

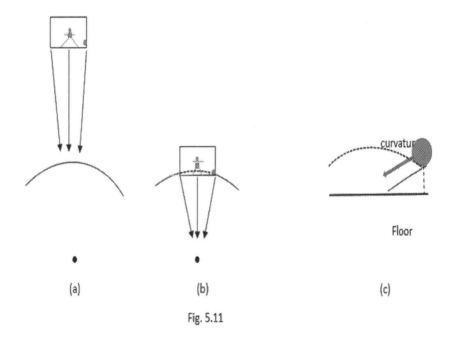

(a)   (b)   (c)

Fig. 5.11

Suppose in Fig 5.5a, reproduced above in Fig. 5.11a, the room is sufficiently large enough that the curvature of the earth would show up, as in Fig 5.11b.

We can see that the curvature of the earth no longer matches the floor, and in this instance, the ball would accelerate towards the center as in Fig. 5.11c which is what we demonstrated in Fig. 5.6c. In everyday life, the floor does match the curvature of the earth because its length is infinitesimal small compared to the curvature of the earth, and thus we are led to think as in the original Galileo's experiment, believing that the ball would continue to move with velocity v along a "straight line".

When Newton proposed his universal Law of Gravity it was met with great skepticism as it was after all a spooky action at a distance. And Einstein often repeated that his GR had explained away this spooky action at a distance. In effect what Einstein proposed can be illustrated schematically as: Gravity → field → energy → matter. It is then that "matter" appears as if it shapes space-time into a curved manifold. Many have interpreted that as replacing "spooky action at a distance" with "space-time is warped". One can debate over which of these two expressions is the least palatable.

5.5 Mass, Inertial or Gravitational

Consider a candle radiating light in all directions.

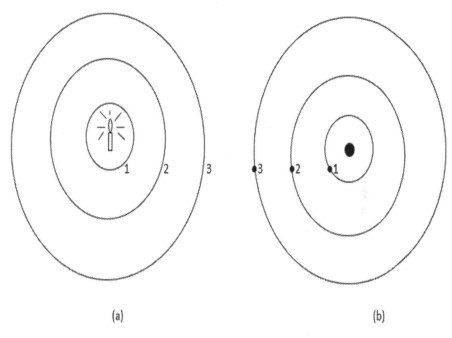

Fig. 5.12

The energy released by the candle spreads out in concentric circles. Since energy is conserved, we find that at point 1 in Fig 5.12a,

$$I_1 = \frac{k}{4\pi R_1^2} \qquad 5.1$$

Where k is related to the energy released at the origin by the candle, and $4\pi R_1^2$ is the surface area of the sphere over which the energy has spread out at R_1 .

Similarly for points 2 and 3,

$$I_2 = \frac{k}{4\pi R_2^2} \quad and \quad I_3 = \frac{k}{4\pi R_3^2} \qquad 5.2$$

In general, for any distance R:

$$I = \frac{k}{4\pi R^2}$$ 5.3

If you would move around circle 1, you would feel the same intensity (energy). Ditto for circles 2 and 3. We have a similar situation with gravity in Fig. 5.12b. Consider a point mass instead of a candle that serves as a source of gravity, and a test mass that we can place on any of the circles 1, 2 or 3.

According to Newton's Law of Gravity,

$$F_g = \frac{GM_{source}M_{test}}{R^2}$$ 5.4

We can define a gravitational field as,

$$g = \frac{F_g}{M_{test}} = \frac{GM_{source}}{R^2}$$ 5.5

We can see that equation 5.5 is similar to equation 5.3.

 Moreover, if you would move around circle 1 in Fig. 5.12b, you would also feel the same energy. These lines are called equipotential, as the potential energy is constant along a given circle. In fact, it would require no energy whatsoever were you to be in motion on any of the equipotential circle around the source of gravity. We can now see that in this description there is no need to refer to perpendicular (tidal) forces as in Fig. 5.6c, nor to a warped space-time as in Fig. 5.11c. In fact this is exactly what Galileo would have discovered, that a particle along a geodesic, which is just an arc on the circle, would move endlessly with velocity v unless acted upon by an external force to change that motion. This is the revised Law of Inertia.

It is no coincidence then that $M_{inertial} = M_{gravitational}$. In fact, it was unnecessary to distinguish an inertial mass from a gravitational mass, as they are not only equivalent, but also they are exactly the same. In the revised Law of Inertia, the mass in Galileo's Inertia Law is a gravitational mass.

5.6 Fields or Energy Levels

As it was mentioned above, around a spherical object, we have equipotential energy levels. Also, we have a source and a test particle in a field. It is understood that the test particle is taken to be so small that its own field compared to the source field is negligible.

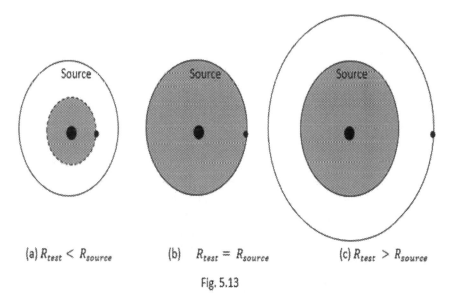

(a) $R_{test} < R_{source}$ (b) $R_{test} = R_{source}$ (c) $R_{test} > R_{source}$

Fig. 5.13

In Fig. 5.13, we have three cases in which,

(i) The test mass is placed inside the source, $R_{test} < R_{source}$

(ii) The test mass is placed on the surface of the source, $R_{test} = R_{source}$.

(iii) The test mass is placed outside the source, $R_{test} > R_{source}$.

In all cases, the test mass will experience a force of gravity as if all the mass inside the circle on which it resides is concentrated at the center. In case (i), the test mass will experience only the force due to the mass inside the dotted line, in other words, a reduced source mass .

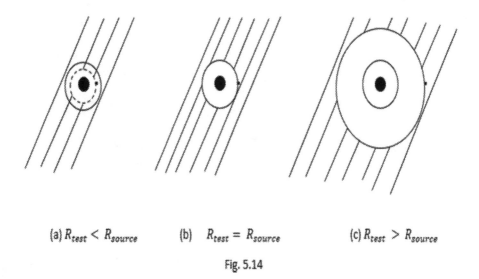

(a) $R_{test} < R_{source}$ (b) $R_{test} = R_{source}$ (c) $R_{test} > R_{source}$

Fig. 5.14

The good news is that we can ignore everything outside the circle of equipotential energy on which the test mass resides, even if the rest is a universe with infinite mass. For the test particle, the only thing that matters is the mass inside the sphere defined by the circle of equipotential energy. We can literally say that gravity lives on a sphere (Fig. 5.14).

5.6.1 Hubble's Law

The important result from the previous discussion is that for any test object that can escape from the pull of gravity from another body (the source) we need to consider only the mass of the source, whether that test particle is a rocket ship or a photon.

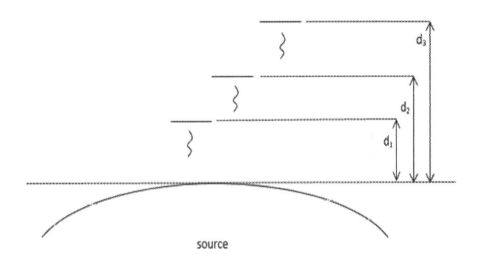

Fig. 5.15

We can now show how Hubble's law can be derived from Newtonian Gravity (NG) and the Equivalence Principle (EP ≡ free falling frame).

Consider this thought experiment. Different emitters are placed at different heights from the ground. The earth plays the role of the source of gravity.

They emit light, which from an earth observer, would be blue-shifted. According to Einstein's Equivalence Principle: we can say that the Doppler Effect is equal to the gravitational shift. [22]

$$(\Delta f/f)_{gravity} = - (\Delta f/f)_{doppler} = -\Delta v/c \qquad 5.6$$

For emitter 1, we can say,

$$\Delta v_1 = g(d_1)\Delta t_1 \qquad 5.7$$

This is just the definition of acceleration, where $g(d_1)$ is the gravitational potential field g at distance d_1.

Define $g(d_1) \equiv g_1$ for simplicity, and using time = distance/velocity, and c is the speed of light.

$$\Delta v_1 = g_1 (d_1/c) \qquad 5.8$$

However from equation 5.5,

$$g_1 = (GM_{source})/ R_1^2 \qquad 5.9$$

$$= (GM_{source}) (R_{source} + d_1)^{-2}$$

$$\approx \{(GM_{source} / R_{source}^2)\} (1 - 2d_1/ R_{source})$$

For $2d_1 << R_{source}$

$$g_1 \approx (GM_{source})/ R_{source}^2 \qquad 5.10$$

Substitute (5.10) into (5.8),

$$\Delta v_1 = \{(GM_{source})/ cR_{source}^2\}(d_1) \qquad 5.11$$

We can get the same result for emitters 2 and 3,

$$\Delta v_2 = \{(GM_{source})/\, cR^2_{source})\}(d_2) \qquad 5.12$$

$$\Delta v_3 = \{(GM_{source})/\, cR^2_{source})\}(d_3) \qquad 5.13$$

We can generalize equations 5.11 to 5.13 as,

$$\Delta v = Hd \qquad 5.14$$

Where $H = (GM_{source})/\, cR^2_{source}$

In case you haven't recognized this, this is Hubble's equation. When Hubble discovered that all galaxies have a redshifted spectrum, Hubble concluded that all the galaxies were moving away. That is the Doppler Effect. However using Einstein's Equivalence Principle, we can say that photons are red-shifted (they are moving against gravity). Note that Hubble discovered not a change in velocity proportional to the distance but just a velocity. In his days, he did not have the technology to observe such a small change in the galaxies' velocities, and it took nearly 70 years before it was discovered that galaxies are actually accelerating. Consider one galaxy against all others (Fig. 5.16).

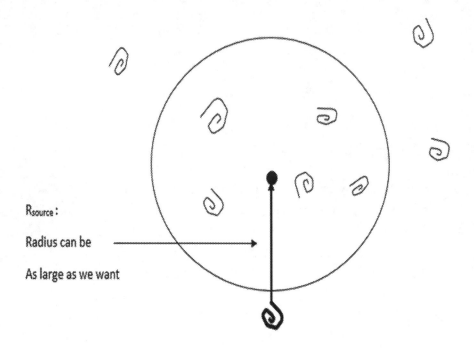

R_{source} :

Radius can be

As large as we want

Fig. 5.16

According to section 5.6, a galaxy would be attracted as if all the matter inside the sphere were concentrated at the center of that sphere. One can ignore all the other galaxies outside that sphere. Any object escaping that galaxy must escape the pull of gravity of all the matter in that sphere, the source. Similarly, we can argue that a photon leaving a galaxy is only influenced by the gravitational pull of the source, which has a very, very large radius in this case.

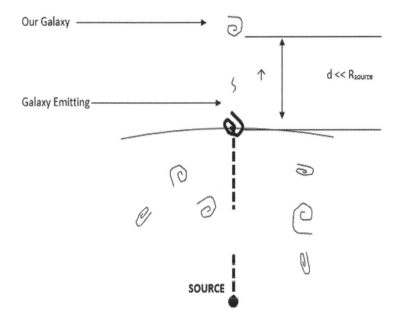

Our Galaxy ⟶

Galaxy Emitting ⟶

$d \ll R_{source}$

SOURCE

Fig. 5.17

Ignoring scales, fig. 5.17 illustrates what is at stakes: Equation 5.14 is true for every galaxy, since each one is emitting light from a source of gravity that has an infinite radius (as large as we want). Needless to say, this argument is only valid if the universe is infinite. So we have two points of view in regard to the observation of the galactic shift of photons from faraway galaxies: (1) a Doppler effect as Hubble interpreted it; (2) a gravitational shift as interpreted from the EP. Hubble's law does not distinguish these two points of view. Furthermore, interpretation (2) yields a universe that could be either accelerating or decelerating.

5.7 Einstein's "Field" Equations

Of course if you insist on working your physics on a local "inertial" flat small patch [23], "inertial" as defined originally by Galileo along a straight line rather than a geodesic, you will experience those weird effects – "there are tidal forces" or "space-time is warped" – as such a patch does not match the curvature of the earth (a spherical object) as demonstrated in Fig. 5.11. In our everyday life, we do not experience these weird effects simply because when we walk around, we are moving along the surface of the earth, an equipotential surface. When we go up an inclined plane, we must exert ourselves to overcome the force of gravity, or when we go down, which it's an easier task as gravity is pulling us down. But we still do not experience these weird effects as our inclined plane is not a flat small patch but rather a small curved surface. Since gravity lives on a sphere, any other kind of coordinate system, especially the flat ones, will produce these weird effects.

Are these effects real? They are as real as the Coriolis force is. Pilots must take into account of the Coriolis force when flying their planes to destination; similarly tidal forces explain the high and low tides of the oceans. But no one dreams of quantizing the Coriolis force. We'll leave to others to decide if quantizing gravity is a worthwhile endeavor.

But now let us examine closely what are the Einstein's "field" equations and what do they really mean.

Einstein reasoned that if "space-time" is curved, then equation 2.19 must be amended,

$$\eta_{\mu\nu} \rightarrow g_{\mu\nu}$$

$$ds^2 = g_{\mu\nu} \, dx^\mu dx^\nu \qquad 5.15$$

Einstein identified [24][25] (more to say in the next section),

$$g_{00} \rightarrow -(1 + \frac{2\varphi}{c^2}) \qquad 5.16$$

Where φ is the gravitational energy potential, and not the field, which is:

$$\varphi = -\frac{GM_{source}}{R} \qquad 5.17$$

It's not a coincidence that the first solution proposed by Schwarzschild for a massive "spherically-symmetric" object yields the Schwarzschild radius as,

$$R_{Schwarzschild} = \frac{2GM}{c^2} \qquad 5.18$$

Then from Newton's law expressed in terms of Gauss' theorem,

$$\nabla^2\varphi = 4\pi\rho \qquad 5.19$$

Where ρ is the mass density, Einstein identified,

$$T_{00} \rightarrow \rho \qquad 5.20$$

Where T_{00} is the energy-momentum tensor. The rest is history. The Einstein "Field" equations are then,

$$R_{\mu\nu} - \tfrac{1}{2}g_{\mu\nu}R = \frac{8\pi G}{c^4} T_{\mu\nu} \qquad 5.21$$

Disregarding the factor in front of the energy-momentum tensor, we see that the Einstein's "Field" equations have a lot more to do with energy than a field. It stands to reason that objects want to follow the path of a geodesic, because that's where equipotential circles exist. It's where planets establish their stable orbits, it's what we as humans do when we walk around effortless on the surface of the earth, and it's where a gravitational mass will move endlessly with velocity v unless an external force compels it to do otherwise. If we insists on describing the laws of physics on a flat coordinate system moving along a "straight" line, which doesn't match a geodesic, then the weird stuff comes out.

5.8 Weak Gravitational Fields

Often this is referred as the Newtonian limit of GR. Right now, we will demonstrate how Einstein's identification came about (equation 5.16).

We will derive the next two equations in section 6.6, equations 6.53 and 6.65, which we will quote below. The first equation is the equation of a geodesic. The second equation is about the Γ's, the affine connection, related to the metric tensor in 5.16.

$$\frac{d^2 x^\lambda}{d\tau^2} + \Gamma^\lambda_{\mu\nu} \frac{dx^\mu}{d\tau} \frac{dx^\nu}{d\tau} = 0 \qquad 5.22$$

$$\Gamma^\lambda_{\mu\nu} = \frac{1}{2} g^{\lambda\rho} \left\{ \frac{\partial g_{\nu\rho}}{\partial x^\mu} + \frac{\partial g_{\mu\rho}}{\partial x^\nu} - \frac{\partial g_{\mu\nu}}{\partial x^\rho} \right\} \qquad 5.23$$

Recall the special free falling frame in fig. 5.3. In that frame, we will now consider a particle moving slowly in a "weak stationary gravitational field". Also, we consider the test particle's velocity to be much smaller than the speed of light, v << c. That is for μ, v = 1,2,3, which are the components of the velocity v^i, the second term in equation 5.22 can be neglected, except for μ, v = 0. From equation 5.22, we are left with $(x^0 = t)$,

$$\frac{d^2x^\lambda}{d\tau^2} + \Gamma^\lambda_{00}\frac{dx^0}{d\tau}\frac{dx^0}{d\tau} = \frac{d^2x^\lambda}{d\tau^2} + \Gamma^\lambda_{00}(\frac{dt}{d\tau})^2 = 0 \quad 5.24$$

And from equation 5.23

$$\Gamma^\lambda_{00} = \frac{1}{2} g^{\lambda\rho} \left\{\frac{\partial g_{0\rho}}{\partial x^0} + \frac{\partial g_{0\rho}}{\partial x^0} - \frac{\partial g_{00}}{\partial x^\rho}\right\}$$

$$= \frac{1}{2} g^{\lambda\rho} \left\{\frac{\partial g_{0\rho}}{\partial t} + \frac{\partial g_{0\rho}}{\partial t} - \frac{\partial g_{00}}{\partial x^\rho}\right\} \quad 5.25$$

To the observer in that special free falling frame, the gravitational field is stationary, and so the derivative of the field with respect to time is zero.

Therefore, equation 5.25 reduces furthermore to,

$$\Gamma^\lambda_{00} = -\frac{1}{2} g^{\lambda\rho} \frac{\partial g_{00}}{\partial x^\rho} \quad 5.26$$

Because the field is weak, we can write the metric tensor as,

$$g_{\lambda\rho} = \eta_{\lambda\rho} + h_{\lambda\rho}, \quad |h_{\lambda\rho}| \ll 1 \quad 5.27$$

$$\Gamma^\lambda_{00} = -\frac{1}{2} \eta^{\lambda\rho} \frac{\partial h_{00}}{\partial x^\rho} \quad 5.28$$

The equation of motion (5.24) becomes for $\lambda, \rho = 1, 2, 3 = i$,

$$\frac{d^2x^i}{d\tau^2} = \frac{1}{2} (\frac{dt}{d\tau})^2 \frac{\partial h_{00}}{\partial x^i} \quad 5.29$$

$$\frac{d^2t}{d\tau^2} = 0 \text{ for } \lambda, \rho = 0 \quad 5.30$$

The second equation (5.30) tells us that $dt/d\tau$ is a constant and so we can divide the first equation (5.29) by $(dt/d\tau)^2$, we get (bold-face for vectors),

$$\frac{d^2\mathbf{x}}{dt^2} = \frac{1}{2} \nabla h_{00} \qquad \text{5.31}$$

The corresponding Newtonian result is,

$$\frac{d^2\mathbf{x}}{dt^2} = -\nabla\varphi \qquad \text{5.32}$$

Where φ is the gravitational potential at a distance r from the center of a body of mass M,

$$\varphi = -\frac{GM}{r} \qquad \text{5.33}$$

Equating 5.31 and 5.32, we get

$$h_{00} = -2\varphi + \text{constant} \qquad \text{5.34}$$

From equation 5.27, Einstein deduced,

$$g_{00} = -(1 + \frac{2\varphi}{c^2}) \qquad \text{5.35}$$

And that is equation 5.24.

The reason this is brought up is to point out why this free falling frame is so special. Nowhere else can one make the claim that the gravitational field is stationary except for the observer in this special frame. Any other observer, whether stranded standing on the surface of a planet, or enjoying a ride on a merry-go-round (non-inertial frame), would claim that the test particle is experiencing a real physical field that is changing while moving through that field. As we shall see later this has significant consequences, particularly when it comes to Gauge theory. The upshot is that observers grounded by gravity on a planet can still consider their frame to be equivalent to an inertial frame and discover all the laws of nature unimpeded by their unfortunate situation.

5.9 Gravitational Time Dilation

In section 2.4, we worked out the time dilation in the case of a body moving at constant speed, in which we found that a moving clock slows down. In the Twin Paradox, Alice who travels away will age slower than her twin Bob.

We find a similar situation if our twins are in different gravitational fields.

Suppose our twins are in a rocket which takes off, accelerating against gravity. Bob is in the tail of the rocket, while Alice is in the nose of the rocket, a distance h from her brother. In terms of the gravitational field, Bob being closer to the center of the earth is in a higher gravitational field than Alice.

We can describe their two positions as:

(i) For Bob,

$$z_B[t] = \tfrac{1}{2}\, gt^2 \qquad\qquad 5.36$$

(ii) For Alice,

$$z_A[t] = h + \tfrac{1}{2}\, gt^2 \qquad\qquad 5.37$$

Note: here, $z[t]$ means that z is a function of t.

Suppose now that Alice sends a light impulse towards Bob. Consider the time when she sends the signal is $t = 0$, and Bob receiving it at $t = t_1$. Therefore we have,

$$z_B[t_1] - z_A[0] = -c(t_1 - 0) \qquad 5.38$$

Recall that going up is positive, so light traveling down carries a negative sign ($-c$). A moment later Δt_A, Alice sends a second light impulse towards Bob. The light will travel the same distance (h) in the same time (t_1) and Bob will receive it at some other time, $t_1 + \Delta t_B$. For the second signal we have,

$$z_B[t_1 + \Delta t_B] - z_A[\Delta t_A] = -c((t_1 + \Delta t_B) - \Delta t_A) \quad 5.39$$

We will now substitute equations (5.36-7) into the above. First in equation 5.38,

$$\tfrac{1}{2} g t_1{}^2 - h = -ct_1 \qquad 5.40$$

Second in equation 5.39,

$$\tfrac{1}{2} g(t_1 + \Delta t_B)^2 - (h + \tfrac{1}{2} g(\Delta t_A)^2)$$
$$= -c(t_1 + \Delta t_B - \Delta t_A) \quad 5.41$$

We will square the bracket in the first term of the LHS.

$$\text{LHS} = \tfrac{1}{2} g(t_1 + \Delta t_B)^2 - (h + \tfrac{1}{2} g(\Delta t_A)^2)$$
$$= \tfrac{1}{2} g(t_1{}^2 + 2t_1\Delta t_B + (\Delta t_B)^2)$$
$$-(h + \tfrac{1}{2} g(\Delta t_A)^2)$$

Consider that we are in a weak gravitational field (section 5.8) we keep only linear terms in Δt_A and Δt_B.

$$\text{LHS} \approx \tfrac{1}{2} g t_1{}^2 + g t_1 \Delta t_B - h$$

Equation 5.41 now reads as,

$$\tfrac{1}{2} g t_1{}^2 + g t_1 \Delta t_B - h = -ct_1 - c\Delta t_B + c\Delta t_A \qquad 5.42$$

Subtract equation 5.40, we get

$$gt_1 \Delta t_B = -c\Delta t_B + c\Delta t_A$$

Or,

$$(gt_1 + c)\Delta t_B = c\Delta t_A \qquad\qquad 5.43$$

Now consider that t_1 is the time taken by light to travel the distance between Alice and Bob, which is h. Hence we have $t_1 = h/c$. Substitute that in the above equation, we get,

$$\left(\frac{gh}{c} + c\right)\Delta t_B = c\Delta t_A \qquad\qquad 5.44$$

Divide both sides by c,

$$\left(\frac{gh}{c^2} + 1\right)\Delta t_B = \Delta t_A \qquad\qquad 5.45$$

Rearranging,

$$\Delta t_B = \frac{\Delta t_A}{\left(1 + \frac{gh}{c^2}\right)}$$

$$= \Delta t_A \left(1 + \frac{gh}{c^2}\right)^{-1}$$

$$\approx \Delta t_A \left(1 - \frac{gh}{c^2}\right)$$

We see that for Bob in a higher gravitational field his clock (Δt_B) is slower than Alice's. With respect to an observer on the ground, the Global Position Satellite (GPS) is moving (its clock slows down) and is in a lower gravitational field (its clock speeds up). Both effects must be taken into consideration in order to synchronize the GPS clock with the observer's clock on the ground.

The Yang-Mills theory revived the old idea that elementary particles might have new degrees of freedom in some kind of "internal" space.

- K Moriyasu[26]

Chapter 6

New Insights into How GR Fails in Gauge Theory

Preliminary: a bird's eye view of Gauge Theory

In this section we will explore the area in which GR crosses lines with QM. But before we tackle this task, we will make a detour into the land of what is known as gauge theory. Initially, gauge theory was about arbitrariness, followed closely by the notion of degrees of freedom. Presently, we believe it's about internal symmetry.

Case A: Consider a system of two equations, with three unknowns x, y, and z. What you get is a redundancy in the values of, say z.

Example: $x + 2y - z = 10$, $3x + 4y + z = 20$ → $z = 5 - x - y$

The values of z can take an infinite number of values.

By adding conditions to x and y we can narrow the values of z. For instance, suppose x, y and z are also functions of time, t. If we have initial conditions such that $x(0) = 2$ and $y(0) = 1$. Then we know that $z(0) = 2$. In other words, adding more information to our system of equations reduces the redundancy.

Case B: A similar situation exists with the Maxwell's equations, which governs the electric field (E) and the magnetic field (B).

There is a redundancy as one can recast E → E + f(x), B → B + g(x) and get the same solution. In other words, just like z in case A, the electromagnetic field can take many values for the same set of equations.

Enters the concept of internal symmetry: microscopic particles like electrons and protons have some internal structure, unavailable to our observations. Through the Langangian, a symmetry means a conservation law (Noether's theorem). Those particles do obey certain conservation laws. Hence we can conclude that whatever that internal structure is, it contains some symmetry, and they can be the source of the needed additional information.

Fortunately, in our math toolkit we have exactly what we need to handle symmetries, and that is Group theory.

So the task is one of identifying the appropriate group for each of the fundamental forces.

This detour begins with the advantages of starting a physical theory with the Lagrangian. The payoff will be a greater understanding of where physics stands today in regards to the unification of all the fundamental forces in nature.

Historically Newton put physics on the map with his three laws of motion, notably the 2nd law, $\mathbf{F} = m\mathbf{a}$. Here we note that we are dealing with vectors. On the other hand the

Lagrangian as a difference between kinetic energy and potential energy (section 2.5) is a scalar, an easier task as we are not bothered with painstaking care of directions as is the case with vectors. But a more advantageous reason is that the Lagrangian is closely related to the Action Principle and symmetry laws.

6.1 Lagrangian and Fields

In the theory of classical fields, the considerations in section 2.5 are repeated with the following changes:

$$x \rightarrow \varphi(x) \qquad 6.1$$

Where x is understood to be the 4-vector space-time $x^\mu = (ct, x^i)$ (see section 2.1), and $\varphi(x)$ is a field.

Also, the Lagrangian and Hamiltonian are replaced by their densities:

$$L \rightarrow \mathcal{L} \text{ and } H \rightarrow \mathcal{H}$$

With these changes, the action of equation 2.39 is now,

$$S = \int \mathcal{L}(\varphi, \partial_\mu \varphi) \, d^4x \qquad 6.2$$

The Euler-Lagrange equations for classical fields (equation 2.43) become,

$$\frac{\partial \mathcal{L}}{\partial \varphi} - \partial_\mu \left(\frac{\partial \mathcal{L}}{\partial (\partial_\mu \varphi)} \right) = 0 \qquad 6.3$$

The conjugate momentum (equation 2.49) is now,

$$\Pi(x) = \frac{\partial \mathcal{L}}{\partial \dot{\varphi}} \qquad 6.4$$

And the Hamiltonian (equation 2.50),

$$\mathcal{H} = \Pi(x)\dot{\varphi} - \mathcal{L} \qquad 6.5$$

6.2 Lagrangian and Symmetry

As it was mentioned before, there is another very important role for the Lagrangian, which is, if the Lagrangian is invariant under a symmetry transformation, then there is a conservation law associated with that symmetry.

For instance, the laws of physics are independent of the coordinate system chosen to write those laws. Consider the case that one observer chooses the origin of a coordinate system at a given point x, and some other observer choose a different point x'. This is known as the symmetry of space-time. This can be illustrated as,

$$x^\mu \;\to\; x'^\mu = x^\mu + \delta x^\mu = x^\mu + a^\mu \qquad 6.6$$

Where a^μ is small and does not depend on x. The corresponding change in the field φ is,

$$\varphi(x) \;\to\; \varphi(x') = \varphi(x + a) = \varphi(x) + \delta\varphi \qquad 6.7$$

Using Taylor expansion, we can write,

$$\varphi(x + a) = \varphi(x) + a^\nu \partial_\nu \varphi \qquad 6.8$$

This gives,

$$\delta\varphi = a^\nu \partial_\nu \varphi \qquad 6.9$$

Now the Lagrangian $\mathcal{L}(\varphi, \partial_\mu \varphi)$ does not depend explicitly on x. A change in the Lagrangian gives,

$$\delta\mathcal{L} = \frac{\partial\mathcal{L}}{\partial\varphi}\delta\varphi + \frac{\partial\mathcal{L}}{\partial(\partial_\mu\varphi)}\delta(\partial_\mu\varphi) \qquad 6.10$$

Using the Euler-Lagrange equations (6.3), we can write the above as a total derivative,

$$\delta\mathcal{L} = \partial_\mu [\left(\frac{\partial\mathcal{L}}{\partial(\partial_\mu\varphi)} \right) \delta\varphi] \qquad 6.11$$

Substitute equation 6.9,

$$\delta\mathcal{L} = \partial_\mu [\left(\frac{\partial\mathcal{L}}{\partial(\partial_\mu\varphi)} \right) \partial_\nu\varphi] a^\nu \qquad 6.12$$

We need one more step. Consider that,

$$\partial_\mu\mathcal{L} \equiv \frac{\delta\mathcal{L}}{\delta x^\mu} \qquad 6.13$$

We can write,

$$\delta\mathcal{L} = \frac{\delta\mathcal{L}}{\delta x^\mu} \delta x^\mu = \frac{\delta\mathcal{L}}{\delta x^\mu} a^\mu = \delta_\nu^\mu \frac{\delta\mathcal{L}}{\delta x^\mu} a^\nu = \partial_\mu(\delta_\nu^\mu \mathcal{L}) \, a^\nu \quad 6.14$$

Where δ_ν^μ is the Kronecker delta function:

$$\delta_\nu^\mu = 1 \; for \; \mu = \nu, and \; \delta_\nu^\mu = 0 \; for \; \mu \neq \nu$$

Equating equations 6.12 and 6.14, and considering that a^ν is arbitrary, we get

$$\partial_\mu \left[\frac{\partial\mathcal{L}}{\partial(\partial_\mu\varphi)} \partial_\nu\varphi - \delta_\nu^\mu\mathcal{L} \right] = 0 \qquad 6.15$$

Now it turns out that the term in the square bracket is,

$$\frac{\partial\mathcal{L}}{\partial(\partial_\mu\varphi)} \partial_\nu\varphi - \delta_\nu^\mu\mathcal{L} \equiv T_\nu^\mu \qquad 6.16$$

Where T_ν^μ is the energy-momentum tensor, from which Einstein had made the following mapping: $T_{00} \rightarrow \rho$, (equation 5.20) in constructing his field equations (equation 5.21).

So equation 6.15 reads as,

$$\partial_\mu T_\nu^\mu = 0 \qquad\qquad 6.17$$

Recall that we are dealing with densities. To get to the energy, we must integrate over volume with $\mu = v = 0$, the time component,

$$\int (\partial_0 T_0^0) d^3 x = \frac{\partial}{\partial t} \int T_0^0 \, d^3 x$$

$$= \frac{\partial}{\partial t} \int \rho \, d^3 x = \frac{\partial}{\partial t} E = 0 \qquad 6.18$$

This equation expresses the conservation of the total energy.

Similarly, for $\mu = 0$ and $v = 1,2,3$, these are the components of momentum densities, and so equation 6.17 with integration over volume corresponds to:

$$\int (\partial_0 T_i^0) d^3 x = \frac{\partial}{\partial t} \int P_i \, d^3 x = 0 \qquad 6.19$$

Again, this expresses the conservation of the total momentum. So we reiterate: a symmetry in the Lagrangian – in the above example, space-time symmetry - yields a conservation law (energy and momentum).

6.3 Gauge Invariance and Electromagnetism

Gauge invariance is an essential concept in understanding the fundamental forces of nature. But the best way to illustrate this concept is by examining the Maxwell's equations which historically provided one of the first cases exhibiting gauge invariance. We will present Maxwell's equations slightly different than in most textbooks. In this description, we have an electromagnetic field $A^\mu(x)$, which is used to form the Lagrangian,

$$\mathcal{L} = -\frac{1}{4}(\partial_\mu A_\nu - \partial_\nu A_\mu)(\partial^\mu A^\nu - \partial^\nu A^\mu) + J^\mu_{em}A^\mu \quad 6.20$$

Where J^μ_{em} is the current density. From this, the equations of motion follows as (see section 2.5),

$$\partial^2 A^\nu - \partial^\nu (\partial_\mu A^\mu) = J^\nu_{em} \qquad 6.21$$

Which cover the first two of Maxwell's equations. What was noticed is that if the electromagnetic field $A^\mu(x)$ is changed with an arbitrary function $\alpha(x)$, the above equation of motions are left unchanged.,

$$A_\mu(x) \rightarrow A_\mu(x) - \partial_\mu \alpha(x) \qquad 6.22$$

This is called a gauge transformation.

To show this, consider the left-hand side of equation 6.21, and substitute equation 6.22,

$$LHS = \partial^2 A^\nu - \partial^\nu \left(\partial_\mu A^\mu\right)$$

$$\rightarrow \partial^2 [A^\nu - \partial^\nu \alpha] - \partial^\nu (\partial_\mu [A^\mu - \partial^\mu \alpha])$$

$$\rightarrow \partial^2 A^\nu - \partial^2 \partial^\nu \alpha - \partial^\nu (\partial_\mu A^\mu - \partial_\mu \partial^\mu \alpha)$$

$$\rightarrow \partial^2 A^\nu - \partial^2 \partial^\nu \alpha - \partial^\nu (\partial_\mu A^\mu) + \partial^\nu \partial_\mu \partial^\mu \alpha$$

Using $\partial_\mu \partial^\mu = \partial^2$ and $\partial^2 \partial^\nu = \partial^\nu \partial^2$, the second term cancels the last term, we end up with,

$$\rightarrow \partial^2 A^\nu - \partial^\nu (\partial_\mu A^\mu)$$

$$= J^\nu_{em} \text{ , which is the RHS of equation 6.21}$$

This results is known as gauge invariance. Physically, it means that if A_μ is the electromagnetic field that solves the equations of motions, so does $A_\mu - \partial_\mu \alpha$. At first sight, there seems to be some arbitrariness hidden in Maxwell's equations. However it had been observed through many experimental observations, that light can be polarized linearly or circularly. The electromagnetic field A_μ has four components, but in the case of polarization, two components can be free but the other two are then fixed. This process is known as gauge fixing. There are many common gauges used in electromagnetism. One of which is the Lorenz gauge,

$$\partial_\mu A^\mu = 0 \qquad\qquad 6.24$$

In the absence of a current density, $J^\nu_{em} = 0$, combining this with the Lorenz gauge in equation 6.21, we get,

$$\partial^2 A^\mu = 0 \qquad\qquad 6.25$$

The solutions are then plane waves of the form,

$$A^\mu = \varepsilon^\mu(p)e^{-ip\cdot x} \qquad\qquad 6.26$$

Where the ε^μ are the plane polarization vectors. The Lorenz gauge has the quality of reducing the number of independent components of the electromagnetic field by one. But that doesn't make A^μ unique. With a second choice, $\nabla \cdot \mathbf{A} = 0$, known as the Coulomb gauge, this further reduces the number of independent components by one. Hence, though the electromagnetic field has four components, the two gauge conditions mentioned above allows only two independent components. We say that the electromagnetic field has two degrees of freedom.

6.4 Gauge Theory and Quantum Field Theory

We will now look at Gauge Theory from a different perspective. In order to do that we must go on some exploration into Quantum Field Theory (QFT).

Consider a complex field defined in terms of two real fields φ_1 and φ_2:

$$\varphi = \frac{1}{\sqrt{2}} (\varphi_1 + i\varphi_2) \qquad 6.27$$

$$\varphi^\dagger = \frac{1}{\sqrt{2}} (\varphi_1 - i\varphi_2) \qquad 6.28$$

For a complex scalar field, the Lagrangian is given as:

$$\mathcal{L} = (\partial^\mu \varphi)^\dagger (\partial_\mu \varphi) - m^2 \varphi^\dagger \varphi \qquad 6.29$$

If you follow sections 6.2 and 6.3, with the above Lagrangian (equation 6.29, and $\varphi^\dagger = \varphi$), you get back the K-G equation for a real scalar field (equation 4.28).

Now this Lagrangian has a gauge symmetry. Consider the following gauge transformations,

$$\varphi = e^{i\alpha} \varphi \qquad 6.30$$

$$\varphi^\dagger = e^{-i\alpha} \varphi^\dagger \qquad 6.31$$

Such transformations are rotations in the complex plane.

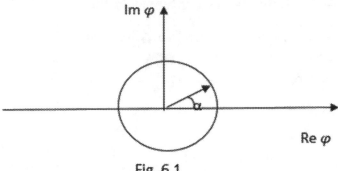

Fig. 6.1

The above Lagrangian is said to be invariant under rotations (Fig. 6.1). And the mathematical framework for this symmetry is Group theory (see the appendix). In this instance we have a U(1) symmetry. The power of gauge theory in QFT is that it opens the door to the richness of Group theory.

Another important point is that equations 6.30-31 are global transformations: they change the field $\varphi(x)$ by the same amount at every space-time point x. But a more important point is to demand local symmetry so that α will have different values at different space-time points. That is,

$$\alpha \rightarrow \alpha(x)$$

$$\varphi(x) = e^{i\alpha(x)} \varphi(x) \qquad 6.32$$

The good news is that the term $m^2\varphi^\dagger\varphi$ in the K-G Lagrangian doesn't change under this new transformations. However, the bad news is that the first term containing derivatives will change. Consider first the following:

$$\partial_\mu \varphi(x) \ \rightarrow \ \partial_\mu[e^{i\alpha(x)} \varphi(x)]$$

$$= e^{i\alpha(x)}\partial_\mu \varphi(x)$$

$$+ e^{i\alpha(x)}i\partial_\mu \alpha(x)\varphi(x)$$

$$= e^{i\alpha(x)}[\partial_\mu + i\partial_\mu \alpha(x)]\varphi(x) \quad 6.33$$

Similarly,

$$(\partial^\mu \varphi(x))^\dagger = e^{-i\alpha(x)}[\partial_\mu - i\partial_\mu \alpha(x)]\varphi^\dagger(x) \quad 6.34$$

The first term in the Lagrangian (equation 6.29) will then transform as:

$$(\partial^\mu \varphi)^\dagger (\partial_\mu \varphi) \ \rightarrow \ (\partial^\mu \varphi)^\dagger (\partial_\mu \varphi) - i(\partial^\mu \alpha)\varphi^\dagger(\partial_\mu \varphi)$$

$$+ i(\partial^\mu \varphi)^\dagger(\partial_\mu \alpha)\varphi + (\partial^\mu \alpha)(\partial_\mu \alpha)\varphi^\dagger \varphi \quad 6.35$$

We no longer have an invariant under a local U(1) transformation. Nevertheless we can still restore this local symmetry by introducing a gauge field $A^\mu(x)$ via a new object called the covariant derivative D_μ. So now, our local gauge theory will contain the following gauge transformations:

$$\varphi(x) \ \rightarrow \ e^{i\alpha(x)} \varphi(x)$$

$$A_\mu(x) \ \rightarrow \ A_\mu(x) - \frac{1}{q}\partial_\mu \alpha(x)$$

$$D_\mu \ \rightarrow \ \partial_\mu + iqA_\mu(x) \quad 6.36$$

The parameter q is the coupling constant that will tell us how strongly the gauge field will interact with other fields.

We can verify the above result quickly.

First transform the derivative,

$$(\partial_\mu \varphi) \rightarrow D_\mu \varphi \rightarrow [\partial_\mu + iqA_\mu]\varphi$$

Next the field φ,

$$\rightarrow [\partial_\mu + iqA_\mu]e^{i\alpha(x)}\varphi$$

Then the gauge field A_μ,

$$\rightarrow \left[\partial_\mu + iq\left(A_\mu - \frac{1}{q}\partial_\mu\alpha\right)\right]\left(e^{i\alpha}\varphi\right)$$

Now work out each term,

$$
\begin{aligned}
(\partial_\mu \varphi) &= e^{i\alpha}(\partial_\mu\varphi) + ie^{i\alpha}(\partial_\mu\alpha)\varphi \\
&\quad + iqA_\mu e^{i\alpha}\varphi - i(\partial_\mu\alpha)e^{i\alpha}\varphi \\
&= (\partial_\mu\varphi)e^{i\alpha} + iqA_\mu e^{i\alpha}\varphi \\
&= e^{i\alpha}(\partial_\mu + iqA_\mu)\varphi \\
&= e^{i\alpha}D_\mu\varphi \qquad\qquad\qquad 6.37
\end{aligned}
$$

Similarly,

$$(\partial^\mu \varphi)^\dagger = e^{-i\alpha}(D_\mu\varphi)^\dagger \qquad\qquad 6.38$$

And now the first term in the Lagrangian (equation 6.29) is invariant, and therefore the complete K-G Lagrangian is invariant under these three gauge transformations (equation 6.36).

Our next task is to compare the gauge theory in QFT (call that GT1) and gauge theory in GR (call that GT2). So our discussion for GT1 will be limited to SU(2) and its role for the weak nuclear force (also known as the weak force).

Our goal is to point out the salient features that distinguish GT1 from GT2.

6.5 Non-Abelian Gauge Theory and the Weak Force

Caveat: it is recommended that you do Group Theory in the appendix before proceeding. If not, you can skip to the conclusion of this section and proceed to the next section provided you're inclined to accept the claims made at the end of this section.

In our gauge transformations (equation 6.32), the coefficient in the exponential function $\alpha(x)$ is just a number. Such gauge transformations are Abelian.

In the appendix we made the case that Lie groups open the door to generators (equation A.11), repeated below:

$$X = \frac{\partial g}{\partial \theta}\Big|_{\theta=0} \qquad\qquad 6.39$$

And a representation of an element can be expressed in terms of the exponential function (equation A.14), reproduced below:

$$D(\theta) = e^{i\theta X} \qquad\qquad 6.40$$

Where the operator X may not necessarily be commutative (non-Abelian). In the U(1) X SU(2) with which we need to deal in the case of the weak force, a general member of this group is (equation A.24),

$$U = e^{-i(\alpha_0 I + \alpha_k \tau_k)} = e^{-i\alpha_0 I} e^{-i\alpha_k \tau_k} \qquad 6.41$$

Where I is the unit matrix. Note: In the appendix we have σ_k, which is mostly used for spin. Here we use τ_k for isospin. But they are the same Pauli matrices (equation A.20). Recall that the values of k are 1,2,3.

Now we want the element in equation 6.41 to be space-time dependent for a local symmetry,

$$U \rightarrow U(x)$$

Going on our experience with the electromagnetic field in the previous section, we must introduce a vector gauge field $B_\mu \tau^0$ with the following gauge transformations,

$$B_\mu(x) \rightarrow B_\mu(x) + \frac{2}{g_1} \partial_\mu \alpha(x) \qquad 6.42$$

$$i\partial_\mu \rightarrow i\partial_\mu - \frac{g_1}{2} B_\mu(x) \qquad 6.43$$

Note that g_1 is a parameter of the theory, and the factor 2 is a convention. This takes care of the U(1) part of the theory. Next we need to take care of the SU(2) part. Define $U(x) \equiv e^{-i\alpha_k(x)\tau_k}$ for simplicity. Again for each generator τ_k, we need a vector gauge field, $W_\mu(x) = W_\mu^k \tau^k$ such that the gauge transformation is,

$$W_\mu(x) \rightarrow U(x)W_\mu(x) U^\dagger(x) + \frac{2i}{g_2} (\partial_\mu U(x)) U^\dagger(x) \quad 6.44$$

Where again g_2 is another parameter of the theory, and again k = 1,2,3.

Note that this $W_\mu(x)$ takes on the form (equation A.22):

$$W_\mu(x) = \begin{pmatrix} W_\mu^3 & W_\mu^1 - iW_\mu^2 \\ W_\mu^1 + iW_\mu^2 & -W_\mu^3 \end{pmatrix} \qquad 6.45$$

The force mediating particles associated with these gauge fields are as follows:

$$W_{\mu\nu}^{+} = \frac{1}{\sqrt{2}}\left(W_{\mu}^{1} - iW_{\mu}^{2}\right) \qquad 6.46$$

$$W_{\mu\nu}^{-} = \frac{1}{\sqrt{2}}\left(W_{\mu}^{1} + iW_{\mu}^{2}\right) \qquad 6.47$$

$$Z_{\mu} = W_{\mu}^{3}\cos\theta_{W} - B_{\mu}\sin\theta_{W} \qquad 6.48$$

Where θ_{W} is the Weinberg angle given by,

$$\cos\theta_{W} = \frac{g_{2}}{(g_{1}^{2}+g_{2}^{2})^{\frac{1}{2}}} \quad \text{and} \quad \sin\theta_{W} = \frac{g_{1}}{(g_{1}^{2}+g_{2}^{2})^{\frac{1}{2}}} \qquad 6.49$$

Noteworthy to remember in the treatment above: the index μ = 0, 1,2,3 are space-time (or Minkowski), while k = 1,2,3 in equation 6.41, the index of the generators is referred as an internal index.

Subsequently after the theory was published, known as the Weinberg-Salem electro-weak theory, the W^{+}, W^{-} and the Z particles were discovered experimentally.

A similar process exists for the Strong force, and its association with SU(3), though that regime includes such notions as asymptotic freedom and color charge confinement, which are absent in this discussion but not necessary for our conclusion.

Conclusion

So what we have demonstrated so far is that gauge theory in QFT (GT1) has a rich mathematical structure beginning with the gauge transformations that includes an exponential function. This in term leads to Lie groups and generators, the latter associated with the particles mediating the fundamental forces:

(i) *Electromagnetic force associated with $U(1)$*,

and its mediating particles: the photons.

(ii) Electroweak force associated with $U(1)XSU(2)$,

and its three mediating particles: W^+, W^- and Z.

(iii) Strong force associated with $SU(3)$,

and its mediating particles: the gluons.

We shall see in the next section that the gauge theory in GR (GT2) not only lacks this rich mathematical structure but is also plagued by another complication.

6.6 Gauge Theory and General Relativity

The Equivalent Principle (EP) states that it is always possible to choose a "locally inertial frame" (LIF) in which gravity is removed. What we have then is a Cartesian frame, which basically allows us to use all the laws of Special Relativity (SR).

So we denote ξ^α as the coordinate system of the free falling frame, and x^μ as any arbitrary coordinate system, often labelled as the "Lab frame". Suppose we have no force acting on a particle inside our free falling frame ($f^\lambda = 0$). We then have no acceleration,

$$\frac{d^2\xi^\lambda}{d\tau^2} = 0 \qquad\qquad 6.50$$

What happens in the Lab? We can safely say that the ξ^α coordinates are functions of the x^μ coordinate system.

$$\xi^\lambda \;\rightarrow\; \xi^\lambda(x^\mu) \qquad\qquad 6.51$$

Therefore,

$$d\xi^{\lambda} = \frac{\partial \xi^{\lambda}}{\partial x^{\mu}} dx^{\mu} \qquad\qquad 6.52$$

And

$$\frac{d\xi^{\lambda}}{d\tau} = \frac{\partial \xi^{\lambda}}{\partial x^{\mu}} \frac{dx^{\mu}}{d\tau}$$

Taking a second derivative (recall 6.50)

$$\frac{d^2\xi^{\lambda}}{d\tau^2} = 0 = \frac{d}{d\tau}\left(\frac{d\xi^{\lambda}}{d\tau}\right) = \frac{d}{d\tau}\left(\frac{\partial \xi^{\lambda}}{\partial x^{\mu}} \frac{dx^{\mu}}{d\tau}\right)$$

Using the Leibniz rule we have,

$$0 = \frac{\partial \xi^{\lambda}}{\partial x^{\mu}} \frac{d^2 x^{\mu}}{d\tau^2} + \frac{\partial^2 \xi^{\lambda}}{\partial x^{\mu} \partial x^{\nu}} \frac{dx^{\mu}}{d\tau} \frac{dx^{\nu}}{d\tau}$$

Multiply throughout by $\frac{\partial x^{\rho}}{\partial \xi^{\lambda}}$, and using the identity $\frac{\partial x^{\rho}}{\partial x^{\mu}} = \delta^{\rho}_{\mu}$. We get,

$$\frac{d^2 x^{\rho}}{d\tau^2} + \Gamma^{\rho}_{\mu\nu} \frac{dx^{\mu}}{d\tau} \frac{dx^{\nu}}{d\tau} = 0 \qquad\qquad 6.53$$

Where the Christoffel symbols $\Gamma^{\lambda}_{\mu\nu}$ are defined as,

$$\Gamma^{\rho}_{\mu\nu} \equiv \frac{\partial x^{\rho}}{\partial \xi^{\lambda}} \frac{\partial^2 \xi^{\lambda}}{\partial x^{\mu} \partial x^{\nu}} \qquad\qquad 6.54$$

The Christoffel symbols also go by the name of affine connection, and the above equation 6.53 is also known in the mathematical world as the geodesic equation (see section 5.4). So what happens is that the trajectory of a particle in the free falling frame is also the path of a line over a curved manifold.

Note that the Christoffel symbols $\Gamma^{\lambda}_{\mu\nu}$ are functions of both the ξ^{α}'s, the special free falling coordinate system, and the x^{μ}'s, the arbitrary coordinate system.

The next step, which you will find in many textbooks, is to get an equation without the special coordinate ξ^α's. This turns out to be misleading. Now we will follow the main line of this argument.

Consider equation 2.19, reproduced below with the free falling coordinate, after all, the claim according to the EP is that all the laws of SR are valid in this free falling frame.

$$ds^2 = \eta_{\alpha\beta}\, d\xi^\alpha d\xi^\beta \qquad 6.55$$

Substitute equation 6.52,

$$ds^2 = \eta_{\alpha\beta}\, \frac{\partial\xi^\alpha}{\partial x^\mu} dx^\mu\, \frac{\partial\xi^\beta}{\partial x^\nu} dx^\nu$$

$$= \eta_{\alpha\beta}\, \frac{\partial\xi^\alpha}{\partial x^\mu}\frac{\partial\xi^\beta}{\partial x^\nu} dx^\mu dx^\nu$$

We write,

$$ds^2 = g_{\mu\nu} dx^\mu dx^\nu \qquad 6.56$$

Where

$$g_{\mu\nu} \equiv \eta_{\alpha\beta}\, \frac{\partial\xi^\alpha}{\partial x^\mu}\frac{\partial\xi^\beta}{\partial x^\nu} \qquad 6.57$$

And that is the metric tensor mentioned in equation 5.15. Now the reasoning is to find the relationship between the $\Gamma^\lambda_{\mu\nu}$ and $g_{\mu\nu}$, believing that the special ξ^α are gone. But as we shall see, the special ξ^α are not gone, they are just hidden.

So here we go. Take the derivative of 6.57, and apply the Leibniz rule,

$$\frac{\partial g_{\mu\nu}}{\partial x^\lambda} = \eta_{\alpha\beta}\left(\frac{\partial^2\xi^\alpha}{\partial x^\lambda \partial x^\mu}\right)\frac{\partial\xi^\beta}{\partial x^\nu} + \eta_{\alpha\beta}\, \frac{\partial\xi^\alpha}{\partial x^\mu}\left(\frac{\partial^2\xi^\beta}{\partial x^\lambda \partial x^\nu}\right) \quad 6.58$$

Consider in the first term, the factor in the bracket, and using the definition (equation 6.54):

$$\left(\frac{\partial^2 \xi^\alpha}{\partial x^\lambda \partial x^\mu}\right) = \Gamma^\rho_{\lambda\mu} \frac{\partial \xi^\alpha}{\partial x^\rho}$$
6.59

Similarly for the second term, the factor in the bracket is:

$$\left(\frac{\partial^2 \xi^\beta}{\partial x^\lambda \partial x^\nu}\right) = \Gamma^\rho_{\lambda\nu} \frac{\partial \xi^\beta}{\partial x^\rho}$$
6.60

Substitute both of these in equation 6.58, we get:

$$\frac{\partial g_{\mu\nu}}{\partial x^\lambda} = \eta_{\alpha\beta} \Gamma^\rho_{\lambda\mu} \frac{\partial \xi^\alpha}{\partial x^\rho} \frac{\partial \xi^\beta}{\partial x^\nu} + \eta_{\alpha\beta} \Gamma^\rho_{\lambda\nu} \frac{\partial \xi^\alpha}{\partial x^\mu} \frac{\partial \xi^\beta}{\partial x^\rho}$$
6.61

Using the definition in equation 6.57, we have:

$$\frac{\partial g_{\mu\nu}}{\partial x^\lambda} = \Gamma^\rho_{\lambda\mu} g_{\rho\nu} + \Gamma^\rho_{\lambda\nu} g_{\mu\rho}$$
6.62

By permuting the indices, we add two terms and subtract the third one, we get:

$$\frac{\partial g_{\mu\nu}}{\partial x^\lambda} + \frac{\partial g_{\lambda\nu}}{\partial x^\mu} - \frac{\partial g_{\mu\lambda}}{\partial x^\nu} = 2 g_{\rho\nu} \Gamma^\rho_{\lambda\mu}$$
6.63

We define the inverse of the metric tensor as,

$$g^{\mu\sigma} g_{\rho\mu} = \delta^\sigma_\rho$$
6.64

We get,

$$\Gamma^\rho_{\lambda\mu} = \frac{1}{2} g^{\nu\rho} \{\frac{\partial g_{\mu\nu}}{\partial x^\lambda} + \frac{\partial g_{\lambda\nu}}{\partial x^\mu} - \frac{\partial g_{\mu\lambda}}{\partial x^\nu}\}$$
6.65

Now we did accomplish the task of writing the Christoffel symbols solely in terms of the metric tensor, but hidden in the definition of the metric tensor is the special free falling frame ξ^α, equation 6.57. This makes the theory still dependent on the free falling frame.

But the bad news doesn't stop here. We now examine how the gauge transformations fit in with GR.

First let's look on how the affine connections translate from an unprimed coordinate system to a primed one. Starting with an arbitrary vector V^μ,

$$V^{\mu'} = \frac{\partial x^{\mu'}}{\partial x^\mu} V^\mu \qquad\qquad 6.66$$

A convenient rule to remember: If the primed index in the LHS (μ') is the upper index, then the primed index will be also be an upper index on the RHS. Ditto if the primed index in the LHS is the lower index, then it will be also a lower index on the RHS. As always, a repeated index (μ) is a dummy index representing a summation. Secondly, there is a factor for each index of the tensor.

Take the derivative of equation 6.66 with respect to $x^{\lambda'}$ and using the identity,

$$\frac{\partial}{\partial x'} \rightarrow \frac{\partial x}{\partial x'}\frac{\partial}{\partial x} \qquad\qquad 6.67$$

We get,

$$\frac{\partial V^{\mu'}}{\partial x^{\lambda'}} = \frac{\partial x^\rho}{\partial x^{\lambda'}}\frac{\partial}{\partial x^\rho}\left(\frac{\partial x^{\mu'}}{\partial x^\mu} V^\mu\right) \qquad\qquad 6.68$$

$$= \frac{\partial x^\rho}{\partial x^{\lambda'}}\frac{\partial x^{\mu'}}{\partial x^\mu}\frac{\partial V^\mu}{\partial x^\rho} + \frac{\partial^2 x^{\mu'}}{\partial x^\rho \partial x^\mu}\frac{\partial x^\rho}{\partial x^{\lambda'}} V^\mu \quad 6.69$$

The last term is extra. To make this gauge transformation invariant, we must look at how the affine connection will transform (equation 6.54), reproduced below,

$$\Gamma^\lambda_{\mu\nu} \equiv \frac{\partial x^\lambda}{\partial \xi^\alpha}\frac{\partial^2 \xi^\alpha}{\partial x^\mu \partial x^\nu} \qquad\qquad 6.70$$

In the primed one, we have,

$$\Gamma^{\lambda'}_{\mu'\nu'} = \frac{\partial x^{\lambda'}}{\partial \xi^\alpha} \frac{\partial^2 \xi^\alpha}{\partial x^{\mu'} \partial x^{\nu'}} = \frac{\partial x^{\lambda'}}{\partial \xi^\alpha} \left(\frac{\partial}{\partial x^{\mu'}}\right)\left(\frac{\partial}{\partial x^{\nu'}}\right) \xi^\alpha$$

Apply the identity (6.67) to each bracket, one at a time,

$$= \frac{\partial x^{\lambda'}}{\partial x^\rho} \frac{\partial x^\rho}{\partial \xi^\alpha} \left[\left(\frac{\partial x^\tau}{\partial x^{\mu'}} \frac{\partial}{\partial x^\tau}\right)\left(\frac{\partial x^\sigma}{\partial x^{\nu'}} \frac{\partial}{\partial x^\sigma}\right)\right] \xi^\alpha$$

$$= \frac{\partial x^{\lambda'}}{\partial x^\rho} \frac{\partial x^\rho}{\partial \xi^\alpha} \left[\frac{\partial x^\tau}{\partial x^{\mu'}} \frac{\partial}{\partial x^\tau} \left(\frac{\partial x^\sigma}{\partial x^{\nu'}} \frac{\partial \xi^\alpha}{\partial x^\sigma}\right)\right]$$

And the Leibniz rule,

$$= \frac{\partial x^{\lambda'}}{\partial x^\rho} \frac{\partial x^\rho}{\partial \xi^\alpha} \left[\frac{\partial x^\tau}{\partial x^{\mu'}} \frac{\partial x^\sigma}{\partial x^{\nu'}} \frac{\partial^2 \xi^\alpha}{\partial x^\tau \partial x^\sigma} + \frac{\partial \xi^\alpha}{\partial x^\sigma} \frac{\partial^2 x^\sigma}{\partial x^{\mu'} \partial x^{\nu'}}\right] \quad 6.71$$

Using the definition of the affine connection, equation 6.70, we get,

$$\Gamma^{\lambda'}_{\mu'\nu'} - \frac{\partial x^{\lambda'}}{\partial x^\rho} \frac{\partial x^\tau}{\partial x^{\mu'}} \frac{\partial x^\sigma}{\partial x^{\nu'}} \Gamma^\rho_{\tau\sigma} + \frac{\partial x^{\lambda'}}{\partial x^\rho} \frac{\partial^2 x^\rho}{\partial x^{\mu'} \partial x^{\nu'}} \quad\quad 6.72$$

To put the above in standard form we need one more step. Consider the identity,

$$\frac{\partial x^{\lambda'}}{\partial x^\rho} \frac{\partial x^\rho}{\partial x^{\nu'}} = \delta^{\lambda'}_{\nu'}$$

Differentiate both sides with respect to $x^{\mu'}$,

$$\frac{\partial}{\partial x^{\mu'}} \left(\frac{\partial x^{\lambda'}}{\partial x^\rho} \frac{\partial x^\rho}{\partial x^{\nu'}}\right) = \frac{\partial}{\partial x^{\mu'}} \delta^{\lambda'}_{\nu'}$$

The RHS is,

$$\frac{\partial}{\partial x^{\mu'}} \delta^{\lambda'}_{\nu'} = 0$$

The LHS is,

$$\frac{\partial x^{\lambda'}}{\partial x^{\rho}}\frac{\partial^2 x^{\rho}}{\partial x^{\mu'}\partial x^{\nu'}} + \frac{\partial x^{\rho}}{\partial x^{\nu'}}\frac{\partial}{\partial x^{\mu'}}\left(\frac{\partial x^{\lambda'}}{\partial x^{\rho}}\right)$$

Equating both sides,

$$\frac{\partial x^{\lambda'}}{\partial x^{\rho}}\frac{\partial^2 x^{\rho}}{\partial x^{\mu'}\partial x^{\nu'}} + \frac{\partial x^{\rho}}{\partial x^{\nu'}}\frac{\partial}{\partial x^{\mu'}}\left(\frac{\partial x^{\lambda'}}{\partial x^{\rho}}\right) = 0$$

Therefore,

$$\frac{\partial x^{\lambda'}}{\partial x^{\rho}}\frac{\partial^2 x^{\rho}}{\partial x^{\mu'}\partial x^{\nu'}} = -\frac{\partial x^{\rho}}{\partial x^{\nu'}}\frac{\partial x^{\sigma}}{\partial x^{\mu'}}\frac{\partial^2 x^{\lambda'}}{\partial x^{\rho}\partial x^{\sigma}} \qquad 6.73$$

Substitute for the second term in equation 6.72,

$$\Gamma^{\lambda'}_{\mu'\nu'} = \frac{\partial x^{\lambda'}}{\partial x^{\rho}}\frac{x^{\tau}}{\partial x^{\mu'}}\frac{\partial x^{\sigma}}{\partial x^{\nu'}}\Gamma^{\rho}_{\tau\sigma} - \frac{\partial x^{\rho}}{\partial x^{\nu'}}\frac{\partial x^{\sigma}}{\partial x^{\mu'}}\frac{\partial^2 x^{\lambda'}}{\partial x^{\rho}\partial x^{\sigma}} \qquad 6.74$$

The first term is what we would get if the affine connection was a tensor. As we had before in equation 6.69, the second term is extra. And now we take our cue from the covariant derivative in section 6.4, equation 6.36, by adding a term that will cancel the extra term. We define a new covariant derivative. Some textbooks uses the semi-colon convention, others uses the ∇ symbol. That is,

$$V^{\mu}{}_{;\nu} = \frac{\partial V^{\mu}}{\partial x^{\nu}} + \Gamma^{\mu}_{\nu\lambda}V^{\lambda} \qquad 6.75$$

Or

$$\nabla_{\nu}V^{\mu} = \frac{\partial V^{\mu}}{\partial x^{\nu}} + \Gamma^{\mu}_{\nu\lambda}V^{\lambda} \qquad 6.76$$

For lower indices,

$$V_{\mu\,;\nu} = \frac{\partial V_{\mu}}{\partial x^{\nu}} - \Gamma^{\lambda}_{\mu\nu}V_{\lambda} \qquad 6.77$$

$$\nabla_\nu V_\mu = \frac{\partial V_\mu}{\partial x^\nu} - \Gamma^\lambda_{\mu\nu} V_\lambda \qquad\qquad 6.78$$

For our purposes, in line with our gauge transformation of equation 6.36, it suffices to use,

$$\partial_\mu \;\rightarrow\; \partial_\mu - \Gamma^{\cdot}_{\mu\cdot} \qquad\qquad 6.79$$

From the definition $g_{\mu\nu} \equiv \eta_{\alpha\beta} \dfrac{\partial \xi^\alpha}{\partial x^\mu} \dfrac{\partial \xi^\beta}{\partial x^\nu}$ (equation 6.57), the metric tensor is sensitive to the choice of a frame. While the coordinates x^μ are arbitrary – the coordinates ξ^α, the free falling frame of reference, are not. They come from a very unique frame of reference. And since the affine connection in equation 6.70 are defined to be derivatives of the metric tensor (equation 6.65), it is also frame dependent. Moreover, the Γ's are not tensors, meaning that they are not Lorentz invariant – observers in different inertial frames will measure different Γ's.

Also, the gauge theory in QM introduces via group theory the concept of generators. This is characterized by the internal index mentioned above. There is no analogue in GR. Gauge transformation in QM does not arise from any form of coordinate transformation but from a phase change, which plays a major role in defining the shape of the gauge fields.

In the development of physics since the onset of QM, physicists came to believe that the fundamental forces of nature were gauge invariant. In that assumption, it was taken that GR was a gauge theory, but if it is, that gauge theory is drastically different than the gauge theory contained within QFT. For instance in the U(1) case, the gauge transformation

$$\partial_\mu \rightarrow \partial_\mu - iqA_\mu$$

is a transformation in phase space that results from the transformation of $\psi \rightarrow e^{iq\alpha(x)} \psi$. While in GR, the transformation

$$\partial_\mu \rightarrow \partial_\mu - \Gamma^{\cdot}_{\mu \cdot}$$

has an inbuilt frame dependence. The mistake is to think that the latter is also a gauge transformation of the same type as in QFT, which it isn't in spite of its appearance. Obviously, using the Einstein field equations to formulate a Lagrangian, and then plug that in QFT, makes the result frame dependent. A theory that is frame dependent will not yield realistic results especially that QFT is at its core a probability theory.

The irony is that SR was formulated to be frame independent: All the laws of physics are valid in every inertial frame, which is the very first postulate of SR. The Lagrangian in QM which was made to be invariant under a Lorentz transformation can also be made to be invariant under a gauge transformation. While in GR, the Lagrangian under a gauge transformation becomes frame dependent mainly because a specific frame system was chosen: the one in which a = -g, (fig. 5.2).

Note that the free falling frame is ubiquitous in GR. It's in the metric tensor (equation 6.57), in the affine connection (equation 6.70) and in the Einstein Field Equations (equation 5.21). There are no other ways to develop GR than through this free falling frame thought experiment. This frame is unique. And this is a tribute to Einstein placing him in a unique position of developing what is considered as one of the greatest achievement in the advancement of human thought. Whether it was sheer genius or accidental, no matter, without Einstein's thought experiment GR doesn't come to fruition. Needless to say that Einstein took 10 years of his most productive years to develop that unique idea into a full-fledge theory commonly known as GR.

Unfortunately, GR's greatest asset is also its greatest liability. The free falling frame is a major obstacle to its quantization via gauge theory since it inescapably makes it a theory that is frame dependent – a notion that cannot be reconciled with the probabilistic nature of QM. Though the other fundamental forces – electromagnetic, Weak and Strong – arise naturally from the principle of gauge invariance, so far it is not the case for gravity using GR.

What is unavoidable is the question: Does gravity belong to the same category as the other three fundamental forces of nature?

GR cannot be the portal for a quantum theory of gravity as it maps the gravitational field into the metric $g_{\mu\nu}$, which is frame dependent. Newtonian physics is more appropriately equipped as it is frame independent, however it is an incomplete theory as it only describes how object behave in the solar system – that's how Newton derived it, so one must consider this point as a crucial guideline. When it comes to describe the motion of stars in the galaxy, it fails. Newtonian gravity (NG) was derived from Kepler's law, and therefore one can only certify its validity for objects within the solar system. NG does not predict the behavior of stars within the galaxy.

Hence one cannot conclude that GN is the correct theory to describe gravity in its entirety. Today's Dark Matter is only a band aid solution as it is introduced into the theory by hand. The real theory of gravity would have Dark Matter emerged from it naturally, if Dark Matter is the real solution to this dilemma.

Can gravity be the subject of a quantized theory? Perhaps, but we must also ask if there is a need for it. Many believe that quantization of gravity is important when it comes to the Planck scale. First, there is no physical evidence that this scale exists in the real world. It was obtained by Planck while he was musing with different powers of some of the fundamental constants: $\hbar$ for QM, c for SR and G for NG. The question one can ask: why only those three constants? Why not the mass of an electron or its charge. These are also universal constants. So the pertinent question is not whether the Planck scale exists but does it play any significant role in the laws of physics. So even if this scale does exist, in term of energy which is about $\sim 10^{18}$ GeV, the most powerful collider present in the world is the LHC with a capacity of about $\sim 10^4$ GeV, the possibility of humanly achieving the Planck scale is not enviable for many decades, if not centuries. Theories about the Planck scale is like shooting darts in the dark at a target staged a continent away.

Now there is this calculation that shows any probing the Planck scale risk to generate black holes. Let's entertain this calculation for a moment. We will use the equation loosely – meaning we will neglect factors of 2 or 2π since we're aiming for a magnitude of scale. We start with the Compton wavelength,

$$\lambda \sim \frac{\hbar}{Mc}$$ 6.80

What we have is a wavelength λ that is needed to probe at some energy level designated by a mass M.

The radius of a black hole is the Schwarzschild radius,

$$R \sim \frac{GM}{c^2}$$ 6.81

Probing a certain scale means we need a wavelength (energy) approximately equal to the size of what we are probing. That is,

$$\lambda \sim R$$ 6.82

Substituting both equation 6.80-81 we get,

$$\frac{\hbar}{Mc} \sim \frac{GM}{c^2}$$

Solving for the mass, we get,

$$M^2 = \frac{\hbar c}{G} \ or \ M = \sqrt{\frac{\hbar c}{G}}$$

And this is exactly the Planck mass.

So even though such experiment may never take place, our present theories, GR and QM, seem to say that probing at the Planck scale is a futile exercise as the probing itself will turn into black holes, objects that cannot be probed.

Cautionary note: just because one takes an equation from QM (6.80) and another from GR (6.81) and equate them (6.82) doesn't mean it necessarily describes something real.

6.7 Summary

(1) The free falling frame is a special frame in which SR is valid. This extends NG to include SR.

(2) GR (= NG + SR) is the only theory we presently have when dealing with:

   (a) Light interacting with a gravitational field (gravitational lensing).

   (b) Matter interacting with a very strong gravitational field (mercury near the sun).

(3) The quantization of GR is hampered by the presence of the special free falling frame, which makes GR a bad candidate for a gauge theory.

(4) The Complete Classical Theory of Gravity (CCTG) is still to be worked out (Fig. 6.2). We know at least one element missing: Galactic Stellar Motion (GSM). Perhaps with new technology, other pieces of the puzzle will be discovered in future years. Once this theory is elaborated then the question will be, can it be turned into a Quantum Theory of Gravity (QTG)?

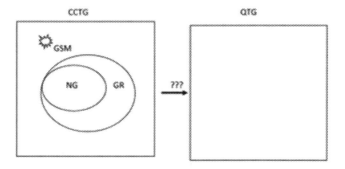

Fig. 6.2

Chapter 7

New Insights into Cosmology

7.1 Olbers' paradox revisited

Olbers' paradox is the argument that the darkness of the night sky conflicts with the assumption of an infinite static universe.

Let n be the average number density of galaxies in the universe. Let L be the average stellar luminosity. The flux f(r) received on earth from a galaxy at a distant r is,

$$f(r) = L/(4\pi r^2) \qquad\qquad 7.1$$

Consider now a thin spherical shell of galaxies of thickness dr. The intensity of radiation from that shell is

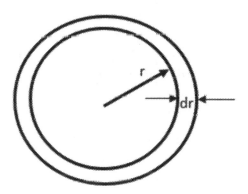

Fig. 7.1

$dJ(r) = $ flux X number of galaxies in the thin shell.

$\qquad = L/(4\pi r^2) \; X \; n \; X \; r^2 dr$

$\qquad = (nL/4\pi)dr \qquad\qquad 7.2$

We can see that the intensity only depends on the thickness of the shell, and not its distance.

The total intensity is found by integrating over shells of all radii.

$$J = (nL/4\pi) \int_0^\infty dr = \infty \qquad 7.3$$

Accordingly, the night sky should be bombarded by an infinite number of photons.

There are two possible explanations readily available (see section 5.6.1):

(1) The Doppler Effect case

The primary argument of the Olbers' paradox from the Big Bang Theory is the universe has a finite age (by extrapolating backward in time), and the galaxies beyond a finite distance, called the horizon distance, are invisible to us simply because they are moving faster than light-speed and therefore their light cannot reach us.

(2) The Gravitational redshift case

The paradox is only an apparent contradiction. In this case, if humans had eyes that could see 7.35 cm wavelength (the Cosmic Microwave Background), then one would see the night sky being illuminated from every direction. What's missing in the above calculation is that the wavelength of light travelling intergalactic distances is shifted more and more towards the red. In terms of the wave model, the next peak would take an infinite amount of time to reach us. This is the surface of infinite redshift. What we see are the photons released from a distance slightly less than the surface of infinite redshift. Any photons released from galaxies beyond that distance will not reach us.

7.2 The Cosmic Microwave Background

In view of what was presented so far in this book we can offer an alternative explanation.

If we accept that gravity is not a Yukawa type of interaction but rather a fictitious force then the Cosmic Microwave Background (CMB) has a simple explanation.

Now let us go back to our probability theory and the example in chapter 4 of rolling a pair of sixes with a pair of even dice, which we said that the odds would be 1/36. If we would find that after 10,000 times rolling the pair of dIce, our observation is 1/36 or very close to that number, no extraordinary explanation would be needed. On the other hand, should we find that our observation of rolling a pair of sixes is, say 10%, which overwhelmingly beats the odds of 1/36 (less than 3%), this would require an explanation other than the simple "all outcome are equally probable" explanation. One of those explanations could be by breaking up the die and find that one of the interior surface is coated with a heavy material. In other words, the inside of the die is not uniformly homogeneous, that is, the die is rigged. The point is: if the probability is 1/36 and the observation is 10% (much greater than the theoretical value), then we need to find more than a simple explanation provided by probability theory.

Similarly, Galileo was concerned that by releasing a 1-kg ball and a 2-kg ball from the same height, at the same time, the two balls reached the ground simultaneously. This seemed to defy the expectation as one would be inclined to think that: (1) the heavy ball should take more time as it would be slower in its motion – something we often observe for moving heavy material; or (2) the heavy ball should take less time as the force of gravity acting on it is greater, hence producing greater acceleration than the light ball. But equal time is a puzzle, which nonetheless was finally resolved by Newton with his three laws of motion and Kepler's laws for orbiting planets yielding an inverse-square law.

What about the CMB?

7.2.1 Thermodynamics Considerations

Consider that photons are primarily created by the stars, they are emitted in every direction. Many are absorbed by nebulae, planets, asteroids and other cosmological objects and re-radiated in every direction. But many are also never absorbed. Considering all cases, many are not only escaping from their source and losing energy, as we have seen in section 5.6.1, they are also distributed randomly in every direction. However that ongoing process of losing energy does not continue infinitely as energy is quantized (chapter 3). That is, at one point, the photon can no longer lose energy: it's either absorbed or not at all. The equilibrium temperature is at, or very near the Absolute Zero temperature. And losing the minimum quantized energy would throw them into negative energy region, which is forbidden by QM considerations. So it is no surprise that the temperature of the CMB is very close to the absolute temperature ($\sim$ 2.7 K), as the photons have the smallest amount of energy permissible without being totally absorbed.

So, shouldn't gravity be able to absorb these minimum energetic photons, in which case there would be no CMB?

7.2.2 QFT Considerations

As we have mentioned in section 4.9, there are two types of particles: fermions (half-integral spins) are the particles of bulk matter; and bosons (integral spins) are the force-mediating particles. If gravity is a Yukawa type of interaction, its theoretical particle, the graviton, would have spin 2 (integral spin) [27], and would now interact with photons, also particles with integral spins. Gluons can interact with themselves as they carry color charge which is necessary for strong interaction. Similarly the W bosons can interact through the electromagnetic force as they carry charge. The Z bosons interact with themselves as they carry mass. The photons however cannot interact through any of these forces as they do not have mass neither do they have charge or color charge. So by these QFT considerations, gravitons are unlikely to interact with photons, and therefore will not absorb the minimum energetic photons of the CMB – these are condemned to roam endlessly in the cosmos.

We can further argue that if gravity is not a Yukawa type of interaction – our initial assumption – and we know that photons necessarily do somehow interact with gravity (section 5.2 Einstein's thought experiment), then the only kind of interaction we can think of would be of the nature of a fictitious force that produce effects like the Coriolis or centrifugal forces - in the case of gravity, it can increase/decrease the photon's energy or deflect their paths. GR does that because at its core, it postulates that in a special frame – the free falling frame - photons behaves as if they follow a curved path on a manifold (gravity is switched off and replaced by an inertial force, section 5.6).

One final note: the CMB is observed as being homogeneous and isotropic. It's a random distribution. If the case had been otherwise, then like the rigged pair of dice, we would need an explanation other than the simple, "photons are emitted in all directions".

7.3 The Case of the Big Bang Theory

As we have argued in section 2.3, the Minkowski diagram of space-time is unreliable as a coordinate system, and should be treated as a graph (Case A) – the so-called "time axis" is really a <u>spatial distance travelled by light</u>. The most reliable mapping that can be considered as a coordinate system, which can depict the real physical world is Case B. Now the Big Bang Theory (BBT) proposes that all matter started from a common point. It is based on Hubble's interpretation that the redshift in spectral analysis of faraway galaxies is a Doppler effect (section 7.1). The closest analogy is that of an explosion occurring at a single point (Fig. 7.2 where a 3-D coordinate system is at play). As different parts of matter are flying apart, "everything is matter moving through space", and given billions of years, (time as in Fig. 1.4 of past, present and future is just a convenient way of depicting the trajectory of the different moving parts of the universe, that is, a parameter), there should be a region of an immense void created from this particular event, a void which would have been increasing in volume.

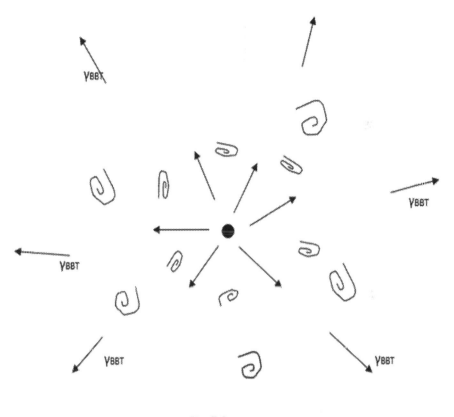

Fig. 7.2

Note that the original photons γ$_{BBT}$ that would be carrying images of the BBT have left our region a long time ago in this model – that is, no CMB. Now such a universe would no longer be homogeneous and isotropic. The fact that there hasn't been any observation of this immense void so far, that there is a CMB, and that the universe is observed as homogeneous and isotropic, constitute strong arguments against the BBT. Needless to say that any cosmological model without a complete theory of gravity is premature, and that includes the above depiction. But the BBT can be defeated using its own arguments.

Postscript: We don't observe a 4-D manifold, even though we can use that mathematical framework to work out our equations. We observe a 3-D world in which time – standardized motion – depicts motion, no more and no less.

7.3.1 The Assumptions of the BBT

Here is a list of assumptions that underlies the BBT – the list is not meant to be exhaustive.

(1) The Equivalence Principle is valid and the redshift is due to a Doppler Effect – the galaxies are moving away from each other.

(2) By extrapolating backward in time, the universe started as a singularity and then expanded.

(3) There is a 4th spatial dimension into which our 3-spatial dimensional world is expanding.

(4) In order to solve the Einstein Field Equations to get to the Friedman Equations, one must assume the universe is homogeneous and isotropic.

(5) To justify (4), one must assume that the universe went through an inflationary period in the early stage.

(6) To justify (5), one must assume that quantum fluctuations popped out of the vacuum some 13.7 billion years.

(7) Since the universe is accelerating, one must assume that the universe is filled with Dark Energy, which must make up 75% of the universe in order to justify a flat space universe (As of now, the Vacuum Energy from (6) is out of step by 122 orders of magnitude with Dark Energy).

(8) To calculate the density of the universe, one must assume the universe is finite in size with its radius equal to its Schwarzschild radius.

(9) In order to tie in the CMB with the BBT, one must assume that the universe must behave like a nearly perfect idealized fluid, so that one can tie in the redshift to the scale factor in (4), which itself is tied in with temperature and time. One can then set a chronology of different reactions that would have happened at different temperatures/times, all of these requiring a number of parameters that can be fine-tuned with observation.

In conclusion, the BBT is a contrived theory which besides the number of assumptions that is needed to support it, nevertheless leaves a certain number of unanswered questions such as:

(a) What evidence do we have that a 4th spatial dimension exists?

(b) If the universe didn't exist from $t = -\infty$ to $t = -13.7$ billion years, what caused it to spring out of the vacuum some 13.7 billion years ago?

(c) How many more assumptions will the BBT need in order to reconcile the Vacuum Energy with Dark Energy in order to make that fit into the theory?

Chapter 8

Why String Theory is Not the Theory of Everything

(The Mathematics of Infinity Can Fool You)

For several decades, String Theory (ST) has had a certain appeal – all the known particles would be made of string of a certain length and frequency. But it has a faulty mathematical foundation. And therefore such a theory is doomed from the start.

One of the most crucial calculations involving ST is the summing all the numbers from 1 to infinity, which intuitively should be infinite. But not in ST, where the sum gives a value of $-1/12$. Yes, if you add all the numbers from 1 to a million, you get a big number; if you add all the numbers from 1 to a trillion, you get an even bigger sum. But should you add all of them to infinity, not only do you get a fraction, but a negative one?! Now I will demonstrate a proof of this result in the first part. It requires to evaluate three infinite sums:

(I) $S_1 = 1 - 1 + 1 - 1 + 1 - ...$

(II) $S_2 = 1 - 2 + 3 - 4 + 5 - ...$

(III) $S_3 = 1 + 2 + 3 + 4 + 5 + ...$

In the second part, I will demonstrate that S_1 gives several different results, therefore invalidating the accepted value of $S_3 = -1/12$, commonly used in String Theory.

8.1 Proof of $S_3 = -1/12$

We start with this infinite series (I):

$$S_1 = 1 - 1 + 1 - 1 + 1 - 1 + ... \qquad 8.1$$

If you pair each term:

$$S_1 = (1 - 1) + (1 - 1) + (1 - 1) ... \qquad 8.2$$

Each pair in the brackets adds to zero, the sum is then

$$S_1 = 0$$

On the other hand, if you pair each one, except the first term:

$$S_1 = 1 + (-1 + 1) + (-1 + 1) + (-1 + 1) ... \qquad 8.3$$

Each pair again adds to zero, but this time,

$$S_1 = 1$$

We have two values for this infinite series. Which is it? So the reasonable answer is take the average, therefore:

$$S_1 = \frac{1}{2}$$

Another way to get the same result is to consider:

$$1 - S_1 = 1 - (1 - 1 + 1 - 1 + 1 - 1 + ... \ )$$

$$= 1 - 1 + 1 - 1 + 1 - 1 + ...$$

$$= S_1$$

Therefore, we again get:

$$S_1 = \tfrac{1}{2} \qquad\qquad 8.4$$

So at this point, one can reasonably accept this result. The next step is to evaluate S_2 (II):

We take two copies of S_2.

$$S_2 = 1 - 2 + 3 - 4 + 5 -\ldots$$

$$S_2 = 1 - 2 + 3 - 4 + 5 -\ldots$$

Where the second line is shifted one column to the right. Now add, term by term:

$$2\,S_2 = 1 - 1 + 1 - 1 + 1 - 1 +\ldots$$

The right-hand side is just equation 8.1, which we established to be equal to $\tfrac{1}{2}$. Therefore,

$$S_2 = 1 - 2 + 3 - 4 + \cdots = \tfrac{1}{4} \qquad\qquad 8.5$$

Now to evaluate S_3 (III), we rewrite it below and subtract S_2 from it:

$$S_3 = 1 + 2 + 3 + 4 + 5 + \ldots$$

$$- S_2 = -1 + 2 - 3 + 4 - 5 +\ldots$$

Term by term, we get:

$$S_3 - S_2 = 4 + 8 + 12 + \ldots$$

We can take out a factor of 4 on the right-hand side, and substitute 8.5 for S_2 in the left-hand side:

$$S_3 - 1/4 = 4 \, (+1 + 2 + 3 + 4 + ...)$$

But now the term in the bracket in the right-hand side is just S_3.

$$S_3 - 1/4 = 4 \, S_3$$

$$S_3 = 1 + 2 + 3 + 4 + 5 + ... = -1/12 \qquad 8.6$$

This is the standard value used in String Theory [28].

8.1.1 The result $S_1 = \frac{1}{2}$ is not unique

Write again equation 8.1,

$$S_1 = 1 - 1 + 1 - 1 + 1 - 1 + ... \qquad 8.7$$

Interchange each term on the right-hand side:

$$S_1 = -1 + 1 - 1 + 1 - 1 + 1... \qquad 8.8$$

We can place brackets as such:

$$S_1 = -1 + (1 - 1 + 1 - 1 + 1...) \qquad 8.9$$

We can add the numbers in the brackets as we did before by considering pairs:

$$S_1 = -1 + \{ (1 - 1) + (1 - 1) + (1 - 1) + ... \} \qquad 8.10$$

Each bracket is zero, therefore

$$S_1 = -1 \qquad\qquad 8.11$$

Another value can also be obtained by taking out a factor of (-1) from the bracket term in equation 8.9:

$$S_1 = -1 + (-1)(-1 + 1 - 1 + 1 - 1 + 1...) \qquad 8.12$$

Using equation 8.8 for the bracket term:

$$S_1 = -1 + (-1)(S_1) \qquad\qquad 8.13$$

Therefore

$$S_1 = -\tfrac{1}{2} \qquad\qquad 8.14$$

In summary, the infinite series $S_1 = 1 - 1 + 1 - 1 + 1 -...$ can take the following values: $0, \pm \tfrac{1}{2}, \pm 1$. This means that $S_3 = 1 + 2 + 3 + ...$ is not unique.

8.2 ST Predicts an Infinite Number of Particles

This doesn't help. Granted that a subset of this infinite number of quantum states can be mapped into the known particles of the standard Model, but so what? ST does not give any new illumination about these known particles. This is also true with the hypothetical graviton: we do get a quantum state with double index polarization vectors, which is what the graviton would need, but then again, no new illumination about the graviton is forked out by ST. That ST contains the graviton is, as it stands right now, just an exaggeration.

8.3 ST needs Supersymmetry

Without Supersymmetry (Susy), ST is a theory of bosons in 26-dimensions. With Susy, it does contain fermions, and the theory is reduced to 11-dimensions. However Susy predicts that for every fermions, there is a partner boson, and vice-versa. So far, none of these extra particles predicted by Susy has surfaced in any experiment.

8.4 Too Many Different Worlds

When attempts are made to reduce this 11-dimensional manifold to our 3+1 world, you then get 10^{500} possible worlds. The only solution is to resign ourselves to the idea that our universe is one of the myriad of possible worlds.

The future for ST does not look very hopeful. First, it is based on a math result that is clearly unreliable. Secondly, it contains extraordinary features that seem to have insurmountable difficulties. But all of this would go away if it could deliver on one single prediction, which it hasn't and seems unlikely of ever doing so.

Epilogue

We have followed a challenging path but a rewarding one. From the very single principle that "everything is matter moving through space", we get our first clue that motion is a fundamental concept, and that time is a mental construct, which enables the measurement of motion (chapter 1); to a revision of what a Minkowski diagram means (chapter 2); to a new law of kinematics with far reaching implications on Hawking radiation (chapter 3); to a sharper interpretation of Quantum Mechanics (chapter 4); to General Relativity and its unsettled conflict with Gauge theory due to the special free falling frame - making gravity behaving more like a fictitious force than a Yukawa type of interacting force - but in spite of this flaw, it is still the only theory available to explain the interaction of light with gravity (chapters 5 and 6); to an explanation of the Cosmic Microwave Background without the need of a cosmological model (chapter 7); and finally to the failures of String Theory.

From a bird's eye view, we are trying to unravel the secrets of the universe using what we can describe as mathematical language, which itself is a play on mappings, and empirical testing as a veracity certification. If you think real hard, this is a crazy idea. But here we are.

Appendix

Group Theory

Group theory plays a fundamental role in physics. One of the main reasons is that it is closely related to symmetry. For instance, the rotation group is related to the fact that the laws of physics are invariant if you rotate your frame of reference.

A.1 Axioms

Group theory is based on four fundamental axioms. These are:

(i) Closure

A group G is a set of elements (a,b,c,...) which include a composition rule such that the product of any two elements is also an element of the group; that is, if a ε G and b ε G then ab ε G.

(ii) Associativity

The composition rule is associative, meaning (ab)c = a(bc)

From axiom (i), a,b,c ε G

Suppose the LHS is: ab = p, and (ab)c = pc = t, where p,t ε G

For the RHS: bc= q, and a(bc) = aq, where q ε G

The associative rule demands that LHS = RHS, and therefore aq = t

 (iii) Identity

The group G has an element e, called the identity element, such that for every element a, ae = ea = a. Note that the identity element for the group is unique.

(iv) Inverse

For every element a ε G there exists an inverse, denoted by a^{-1} such that a a^{-1} = a^{-1}a = e, where again e is the identity element.

A.2 Representation of Groups

A representation is a mapping that takes a group element g ε G into a linear operator F, denoted by,

$$F: g \rightarrow F(G)$$

 Such that the composition rule is preserved, meaning:

(i) Closure: F(a)F(b) = F(ab)

(ii) The identity is preserved: F(e) = I.

The groups G and F(G) are said to be isomorphic, meaning that both groups have the same mathematical structure.

A.3 The Rotation Group

A function can be viewed as given an input x, and the function uses that input to yield an output y:

$$f : x \rightarrow y$$

Or, what is commonly known as y = f(x). Similarly, the individual elements of a group are outputs of function, and the inputs are called parameters.

The rotation group is the set of all rotations about the origin, (Fig. 6.1). In 2-D, the elements contain one-parameter elements obeying the composition rule,

$$R(\theta_1)R(\theta_2) \ = \ R(\theta_1 + \theta_2) \qquad\qquad \text{A.1}$$

The inverse is,

$$R^{-1}(\theta) \ = \ R(-\theta) \qquad\qquad \text{A.2}$$

And the identity element is,

$$I \ = \ R(0) \qquad\qquad \text{A.3}$$

Representation of the rotation group: consider a vector with coordinates x_i. Let x_i' be the coordinates of the vector rotated by an angle θ in the plane. The components of these two vectors are related by the transformation:

$$x_i' \ = \ R_{ij}x_i \qquad\qquad \text{A.4}$$

In this case, the rotation can be represented by a 2-D matrix,

$$R(\theta) \ = \ \begin{pmatrix} cos\theta & sin\theta \\ -sin0 & cos\theta \end{pmatrix} \qquad\qquad \text{A.5}$$

Consider the transpose of this matrix,

$$R^T(\theta) \ = \ \begin{pmatrix} cos\theta & -sin\theta \\ sin\theta & cos\theta \end{pmatrix} \qquad\qquad \text{A.6}$$

Multiply the above matrices,

$$R(\theta)R^T(\theta) \ = \ \begin{pmatrix} cos\theta & sin\theta \\ -sin\theta & cos\theta \end{pmatrix} \begin{pmatrix} cos\theta & -sin\theta \\ sin\theta & cos\theta \end{pmatrix}$$

$$= \begin{pmatrix} cos^2\theta + sin^2\theta & -cos\theta sin\theta + cos\theta sin\theta \\ -sin\theta cos\theta + cos\theta sin\theta & cos^2\theta + sin^2\theta \end{pmatrix}$$

$$= \begin{pmatrix} 1 & 0 \\ 0 & 1 \end{pmatrix} = I \qquad \text{A.7}$$

This means that the transpose is also the inverse,

$$R^{-1}(\theta) = R^T(\theta) \qquad \text{A.8}$$

Another important feature is the determinant,

$$det \ R(\theta) = det \begin{pmatrix} cos\theta & sin\theta \\ -sin\theta & cos\theta \end{pmatrix}$$

$$= cos^2\theta + sin^2\theta = 1 \quad \text{A.9}$$

The 2-D rotation group is part of a larger group, the group $SO(N)$, which are special orthogonal $N \times N$ matrices. They are orthogonal because $R^{-1}(\theta) = R^T(\theta)$ (equation A.8), and special because $det \ R(\theta) = 1$ (equation A.9).

This treatment can be extended to higher dimensions. For example in 3-D, we would need a set of two-parameter elements: $R \rightarrow R(\theta_1, \theta_2)$.

Note that the parameter θ is continuous and so the group SO(N) is said to be a continuous group.

The range of the parameter θ is between 0… 2π. We say that the group SO(N) is compact.

If every elements of the group commutes among themselves, that is, for any elements a ε G and b ε G, we have ab = ba, the group is said to be abelian. If not, the group is said to be non-abelian.

If the number of elements in the group G is finite, the group is said to be finite. Likewise, if the number of elements is infinite, the group is infinite.

A.4 Lie Groups and Generators

Here we confine ourselves to a group G that has the following properties:

(i) There is a finite set of continuous parameters θ_i.

(ii) There exists derivatives of the group elements with respect to all the parameters.

We call this group a Lie group.

One of the key features of Lie groups is the notion of generators. To illustrate this, we will consider a one-parameter element of a Lie group G.

We obtain the identity element by setting $\theta = 0$,

$$g(\theta)|_{\theta=0} = e \qquad \text{A.10}$$

The generators of the Lie group are defined as the derivatives of the group element with respect to the parameter at $\theta = 0$:

$$X = \frac{\partial g}{\partial \theta}\Big|_{\theta=0} \qquad \text{A.11}$$

Where X is the generator. If there are n parameters in the Lie group G, then we have n generators:

$$X_i = \frac{\partial g}{\partial \theta_i}\Big|_{\theta=0} \text{ where } i = 1 \dots n \qquad \text{A.12}$$

Because Hermitian operators $(X^T = X)$ in QM are ubiquitous, we want a unitary representation (see next section) and choose the generators to be Hermitian, that is,

$$X_i = -i \left. \frac{\partial g}{\partial \theta_i} \right|_{\theta=0} \qquad \text{A.13}$$

This unitary representation is best expressed by considering some changes in the parameter θ_i around the origin. We denote the representation by D and express it in terms of the exponential:

$$D(\theta) = e^{i\theta X} \qquad \text{A.14}$$

We can check that it is unitary:

$$D^\dagger(\theta)D(\theta) = e^{-i\theta X} e^{i\theta X} = 1 \qquad \text{A.15}$$

The importance of the generators is that they themselves form a vector space and obey the commutation relation:

$$[X_i, X_j] = i f_{ijk} X_k \qquad \text{A.16}$$

Where f_{ijk} are the structure constants, and the above equation forms the composition rule of a Lie algebra.

A.5 Unitary Groups

Complex numbers are ubiquitous in QM, and to deal with that reality, we need the unitary representations as discussed above. The unitary group U(n) consists of $N \times N$ unitary matrices, for which we have,

$$U^\dagger U = 1 \quad \text{or} \quad U^\dagger = U^{-1} \qquad \text{A.17}$$

The special unitary group is denoted by SU(N), for which det = 1.

The number of generators for a U(N) group is N^2, and for a SU(N) group, $N^2 - 1$.

The simplest unitary group is the group U(1), which is a 1×1 matrix, that is, the complex number,

$$U(\theta) = e^{-i\theta} \qquad \text{A.18}$$

This is the familiar unit circle. We have one generator, $(N^2 = 1^2 = 1)$.

The next unitary group of interest is the SU(2) group with 3 generators ($N^2 - 1 = 2^2 - 1 = 3$). These are related to the Pauli matrices, with the following commutation relationship,

$$\left[\frac{\sigma_i}{2}, \frac{\sigma_j}{2}\right] = i\varepsilon_{ijk}\frac{\sigma_k}{2} \qquad \text{A.19}$$

And the Pauli matrices are,

$$\sigma_1 = \begin{pmatrix} 0 & 1 \\ 1 & 0 \end{pmatrix}, \sigma_2 = \begin{pmatrix} 0 & -i \\ i & 0 \end{pmatrix}, \sigma_3 = \begin{pmatrix} 1 & 0 \\ 0 & -1 \end{pmatrix} \qquad \text{A.20}$$

We can write an element of SU(2) as,

$$U = e^{i\alpha_i\sigma_i/2} \qquad \text{A.21}$$

With α_i being a real number.

Of great interest is the case of the weak nuclear force which is the product of the groups U(1) X SU(2). Consider a Hermitian matrix of the form,

$$H = \begin{pmatrix} \alpha_0 + \alpha_3 & \alpha_1 - i\alpha_2 \\ \alpha_1 + i\alpha_2 & \alpha_0 - \alpha_3 \end{pmatrix} \qquad \text{A.22}$$

We can now split this matrix as,

$$H = \alpha_0 I + \alpha_k \sigma_k \qquad\qquad \text{A.23}$$

Where k now runs from 1 to 3.

A general member of this group is,

$$U = \exp(\alpha_0 I + \alpha_k \sigma_k) = \exp(\alpha_0 I)\exp(\alpha_k \sigma_k) \ \ \text{A.24}$$

The phase factor $\exp(\alpha_0 I)$ belong to a U(1) group (equation A.18), and the second phase factor $\exp(\alpha_k \sigma_k)$ to SU(2) (equation A.21). This element belongs to the product U(1) X SU(2).

References

[1] T. Lancaster, S.J. Blundell; Quantum Field Theory for the Gifted Amateur, Oxford University Press, 2014, page228.

[2] A. EINSTEIN, DOES THE INERTIA OF A BODY DEPEND UPON ITS ENERGY-CONTENT? September 27, 1905 (English translation available at: (http://www.fourmilab.ch/etexts/einstein/E_mc2/e_mc2.pdf).

[3] Joseph Palazzo, A New Law of Kinematics, http://vixra.org/abs/1608.0392, 2016-08-29.

[4] Joseph Palazzo, Are the Laws of Physics Weird? Authorhouse, 2016.

[5] V. Mukhanov, Physical Foundations of Cosmology, Cambridge University Press, 2013, page 35.

[6] F. Reif , Statistical Physics, Berkeley Physics Course – Vol 5, McGraw-Hill Book Company, 1967, page 115.

[7] T. Lancaster, S.J. Blundell; Quantum Field Theory for the Gifted Amateur, Oxford University Press, 2014, chapter 16.

[8] T. Lancaster, S.J. Blundell; Quantum Field Theory for the Gifted Amateur, Oxford University Press, 2014, chapter 17.

[9] S. Hawking, Nature 248, 30 (1974).

[10] S. Hawking, Comm. Math. Phys 43, 199 (1975).

[11] Ta-Pei Cheng, Relativity, Gravitation and Cosmology, Oxford University Press, 2012, page 250.

[12]Leonard Susskind, The Black Hole War, Back Bay Company, 2008. Pp209-210.

[13] Almheiri, Ahmed; Marolf, Donald; Polchinski, Joseph; Sully, James (11 February 2013). "Black holes: complementarity or firewalls?" Journal of High Energy Physics 2013 (2).

[14] J.R. Oppenheimer, H. Snyder; Phys. Rev. 56, 455 (1939).

[15] Matthew D. Schwartz, Quantum Field Theory and the Standard Model, Cambridge University Press, 2014, page 200.

[16] Joseph Palazzo, The Collapse of the Wave Function, http://vixra.org/abs/1608.0350, 2016-08-25.

[17] A. Einstein, "Physics, philosophy and scientific progress", J. Int. Cool. Surg. 14, 755-758 (1950).

[18] J.S. Bell (1964), "On the Einstein-Podolsky-Rosen Paradox", Physics 1: 195–200.

[19]Alain Aspect, Philippe Grangier, Gérard Roger (1982), Phys. Rev. Lett. 49 (2): 91–4.

[20] A. Einstein, B. Podolsky, N Rosen; Physical Review 47(10) 777-780, bicode 1935 PhRv47.77E.

[21] R. Shankar, Principles of Quantum Mechanics, Second Edition, Springer,1994, chapter 8.

[22] Ta-Pei Cheng, Relativity, Gravitation and Cosmology, Oxford University Press, 2012, page 70.

[23] Steven Weinberg, Gravitation and Cosmology, John Wiley & Sons, 1972, pages 19.

[24]James B. Hartle; Gravity, an Introduction to Einstein's General Relativity, Addison-Wesley, 2003, page 189.

[25]Ta-Pei Cheng; Relativity, Gravitation and Cosmology, Oxford University Press, 2012, page 109.

[26] K Moriyasu, An Elementary Primer for Gauge Theory, World Scientific (1983), page 29.

[27] A. Zee, Quantum Field in a Nutshell, Princeton University Press (2003), page 34.

[28] B. Zwiebach, A First Course in String Theory, Cambridge University Press, 2004, page 221.

FROM THE SAME AUTHOR

FICTIONAL

ZOHRA: THE PLANET OF TRUTH AND KNOWLEDGE,
1stBooks, 2003.

ZOHRA II: THE NEW QUEEN, AuthorHouse, 2006.

NONFICTIONAL

ARE THE LAWS OF PHYSICS WEIRD? AuthorHouse, 2016.

Cover painting, "Approaching Storm"

By Rita Palazzo

Printed in the United States
By Bookmasters